NATURE'S PRINCIPLES

LOGIC, EPISTEMOLOGY, AND THE UNITY OF SCIENCE

VOLUME 4

Logic, Epistemology, and the Unity of Science aims to reconsider the question of the unity of science in light of recent developments in logic. At present, no single logical, semantical or methodological framework dominates the philosophy of science. However, the editors of this series believe that formal techniques like, for example, independence friendly logic, dialogical logics, multimodal logics, game theoretic semantics and linear logics, have the potential to cast new light on basic issues in the discussion of the unity of science.
This series provides a venue where philosophers and logicians can apply specific technical insights to fundamental philosophical problems. While the series is open to a wide variety of perspectives, including the study and analysis of argumentation and the critical discussion of the relationship between logic and the philosophy of science, the aim is to provide an integrated picture of the scientific enterprise in all its diversity.

Nature's Principles

Edited by

JAN FAYE
*University of Copenhagen,
Denmark*

PAUL NEEDHAM
*Stockholm University,
Sweden*

UWE SCHEFFLER
*Humboldt University Berlin,
Germany*

and

MAX URCHS
*University of Konstanz,
Germany*

Springer

A C.I.P. Catalogue record for this book is available from the Library of Congress.

ISBN-10 1-4020-3257-9 (HB) Springer Dordrecht, Berlin, Heidelberg, New York
ISBN-10 1-4020-3258-7 (e-book) Springer Dordrecht, Berlin, Heidelberg, New York
ISBN-13 978-1-4020-3257-8 (HB) Springer Dordrecht, Berlin, Heidelberg, New York
ISBN-13 978-1-4020-3258-5 (e-book) Springer Dordrecht, Berlin, Heidelberg, New York

---

Published by Springer,
P.O. Box 17, 3300 AA Dordrecht, The Netherlands.

Cover image:
Adaptation of a Persian astrolabe (brass, 1712-13), from the collection of the Museum of the History of Science, Oxford. Reproduced by permission.

Printed on acid-free paper

Printed in the Netherlands.

Contents

Preface

Most of the present papers were presented at the 5th Baltic Workshop on Logic and Philosophy of Science held at Copenhagen May 24–27, 2001. The workshop carried the title *Language Rules and Laws of Nature* and was made possible by the *Danish Institute for Advanced Studies in the Humanities* and the *Danish Research Council for the Humanities.* We wish to express our gratitude for their generous financial support.

JAN FAYE, PAUL NEEDHAM, UWE SCHEFFLER, MAX URCHS

Contributing Authors

Mauro Dorato is Professor of Philosophy of Science at the University of Rome Three. Among his publications, *Time and Reality. Spacetime Physics and the Objectivity of Temporal Becoming*, (Bologna 1995) and *The Software of the Universe. An introduction to the history and philosophy of the laws of nature*, under translation for Ashgate.

Address: Department of Philosophy
University of Rome Three
Via Ostiense 234, 00146 Roma
dorato@uniroma3.it

Jan Faye is University Professor of Philosophy at Copenhagen University. He is author of *The Reality of the Future* and *Rethinking Science.*

Address: Copenhagen
Njalsgade
faye@hum.ku.dk

Igor Hanzel is Associate Professor of Philosophy at *Comenius University.* He is author of *The Concept of Scientific Law in the Philosophy of Science and Epistemology.*

Address: Dept. of Philosophy of Science
Comenius University
81801 Bratislava, Slovak Republic
HANZEL999@hotmail.com

Henrik Hållsten is currently a visitor at the University of Canterbury, New Zealand, on a STINT postdoctoral scholarship. He wrote his PhD thesis *Explanation and Deduction* (Stockholm 2001) at Stockholm University and has mainly been teaching there since.
Address: Stockholms universitet
106 91 Stockholm
Sweden
henrik.hallsten@philosophy.su.se

Lars-Göran Johansson is Associate Professor of Philosophy at Uppsala University. He is the author of *Understanding Quantum Mechanics. A realist interpretation without hidden variables*, A&W International, 1992 and *Introduktion till Vetenskapsteorin*, Thales, 2003.
Address: Department of Philosophy
Uppsala University
75310 Uppsala, Sweden
lars-goran.johansson@filosofi.uu.se

Geert Keil teaches philosophy at Humboldt University in Berlin. He is the author of *Kritik des Naturalismus* (Berlin/New York 1993), *Handeln und Verursachen* (Frankfurt am Main 2000), *Quine* (Hamburg 2002), and the co-editor of *Fifty Years of Quine's 'Two Dogmas'* (Amsterdam/New York 2003).
Address: Institut für Philosophie
Humboldt-Universität
Unter den Linden 6
D-10099 Berlin, Germany
KeilG@philosophie.hu-berlin.de

Max Kistler teaches philosophy at the University of Paris X in Nanterre, and is member of Institut Jean Nicod, Paris. He is the author of *Causalité et lois de la nature* (Paris, Vrin, 1999), co-author of *La philosophie des sciences au XXe siècle* (Paris, Flammarion, 2000), and co-editor of *Causes et dispositions* (forthcoming, Paris, PUF).
Address: Département de philosophie
Université Paris X – Nanterre
200 avenue de la République
F 92001 Nanterre Cedex, France
kistler@ehess.fr

Uwe Meixner is professor of philosophy at the University of Regensburg. His publications include the recent books *Theorie der Kausalität* (mentis 2001), *The Two Sides of Being* (mentis 2004), and the recent article *Causation in a New Old Key* (Studia Logica 76 (2004)). He is co-editor of two journals: *Logical Analysis and History of Philosophy* and *Metaphysica*.
Address: Institute of Philosophy
University of Regensburg
93040 Regensburg, Germany
uwe.meixner@psk.uni-regensburg.de

Paul Needham is professor of philosophy at the University of Stockholm.
Address: Department of Philosophy
Stockholm University
106 91 Stockholm, Sweden
Paul.Needham@philosophy.su.se

Johannes Persson is senior lecturer in theoretical philosophy and chairman of the department of Philosophy, Lund University. He is the author of *Causal facts* (Stockholm 1997) and interested in problems at the borderline of metaphysics, epistemology, and philosophy of science.
Address: Filosofiska institutionen
Kungshuset, Lundagård
SE-222 22 Lund, Sweden.
johannes.persson@fil.lu.se

Uwe Scheffler teaches philosophy at Humboldt University, Berlin. He is the author of *Ereignis und Zeit* and edited several books on logic and philosophy.
Address: Institut für Philosophie
Humboldt-Universität
Unter den Linden 6
D-10099 Berlin, Germany
uwe.scheffler@alumni.hu-berlin.de

Gerhard Schurz is chair of Theoretical Philosophy at the philosophical department of the university of Duesseldorf (Germany). He is the author of many papers and some books and editions, especially in the field of logic and philosophy of science, including *Wissenschaftliche Erklärung* (Graz 1983), *Erklären und Verstehen in der Wissenschaft* (Oldenbourg, Munich, 1988, 1990), *Law and Prediction in the Light of Chaos Research* (with P. Weingartner, Berlin 1996).
Address: Universität Düsseldorf
Philosophisches Institut,
Lehrstuhl für Theoretische Philosophie
Universitätsstrasse 1, Geb. 23.21
D-40225 Düsseldorf, Deutschland (Germany)
gerhard.schurz@phil-fak.uni-duesseldorf.de

Werner Stelzner works in logic and in history of logic. He has written the book *Epistemische Logik* (1984) and edited the book *Zwischen traditioneller und moderner Logik. Nichtklassische Ansätze* (2001).
Address: Stauffenbergstr. 28
PF 150211
G-7747 Jena, Germany
wstelzner@t-online.de

Max Urchs is professor of philosophy and author of *Klassische Logik* (1993) and *Maschine Körper Geist. Eine Einführung in die Kognitionswissenschaft* (2002).
Address: Universität Konstanz
Fachbereich Philosophie
78464 Konstanz, Germany
max.urchs@uni-konstanz.de

INTRODUCTION

Jan Faye
Paul Needham
Uwe Scheffler
and Max Urchs

1. Preamble

Descartes envisaged an infinite universe in which order could no longer be described, as Aristotle had done, with reference to a unique point at the centre, but depended entirely upon the idea of following universal laws. Newton accepted the general idea, but better appreciated the enormity of the task of actually formulating such laws having found wanting many of Descartes' arguments purporting to subsume phenomena under his mechanical principles. Newton's own attempts led him to a confront the a priori imposition of the principle of action by contact with an array of arguments subsuming a variety of phenomena under his law of gravitation. Because it was unrivalled by any equally articulated theory, the body of scientific opinion soon came to accept this principle of action at a distance, which must surely count as one of the most successful theories of modern science as measured by the time it has reigned unchallenged by any articulate alternative. This is an important point to bear in mind when considering Hume's regularity theory of causation.

Philosophers have also wrestled with the problem of how to articulate the general notion of a law of nature. Their efforts are likewise motivated by obscurities they perceive in certain conceptions, which they try to remedy with more or less detailed frameworks of their own. One leading idea lying at the heart of many subsequent treatments derives from Hume's analysis of causation. Even those who dispute the claims of this tradition must address the *regularity theory*, an exposition of which therefore provides a suitable starting point for this overview of theories of lawlikeness in modern philosophy.

J. Faye, P. Needham, U. Scheffler and M. Urchs, (eds.), Nature's Principles, 1–53.

2. The Regularity Theory of Causation

Intuitively, the concept of causation involves a kind of necessity. We say such things as, given the cause, the effect had to occur, and if it were not for the cause, the effect would not have occurred – expressions which suggest that the cause necessitates its effect. Hume opens his discussion by "observing that the terms *efficacy, agency, power, force, energy, necessary connexion*, and *productive quality*" – several of which were used by Newton – "are all nearly synonymous" (Treatise 1739, I.III.xiv)[1]. No one of them can therefore be used to explain any of the others, and he goes on to consider what independent sense can be made of this notion of necessary connection. The first point he makes is that it cannot be logical necessity – it cannot, as he puts it, be a matter of a "relation of ideas". For the idea of the effect is distinct from the idea of the cause, and reason alone is not sufficient to obtain the idea of the effect from the idea of the cause – "Reason alone can never give rise to any original idea" (Treatise 1739, I.III.xiv). In more modern jargon, Hume's first point was that given a description of the cause, a description of the effect does not follow logically. Accordingly, whatever the connection between cause and effect, it is not an analytic truth that the particular event which happens to be the effect follows the particular event which happens to be the cause. The existence of a causal relation is a substantial fact about the world, and statements of causal connections are synthetic truths.

Having denied that causal connections involve analytic necessity, David Hume's second point is to deny that the causal relation is a necessary relation at all. He argues that if we consider what we see when one billiard ball collides with and causes another to move, all we observe is the movement of the one ball and then the movement of the other. There is no connection to be seen, only the cause and the effect. "All ideas are deriv'd from, and represent impressions. We never have any impression, that contains any power or efficacy. We never therefore have any idea of power" (Treatise 1739, I.III.xiv). He concludes that since there is no such connection to be seen, there is no necessary connection at all.

There is an air of paradox in this. We began by talking about the kind of causal connection we express by saying that, given the cause, the effect had to occur, and conclude that no such connection is observable. But if this putative connection is not observable, how could we recognise and

[1]This manner of referring to Book I, Chapter III, Section xiv of Hume's *Treatise of Human Nature* follows a standard format, independent of the particular edition used here (for which, see the bibliography).

discuss it in the first place? There must be something distinguishable about causally related events since some, but not all, pairs of events are causally related. And if it is not the connection that distinguishes them, what does? At the very least we must be able to explain how the original question of the nature of the causal necessity arose. Hume certainly didn't want to deny the distinction between causally related and non-related events, and the theory he elaborated addressed these questions. Specifically, he distinguished two problems: (i) How is our conviction that the effect *must*, so it seems, follow the cause to be explained? and (ii) What does distinguish causally connected events from others?

Hume answers the first question with the claim that

> Necessity is something that exists in the mind, not in objects (Treatise 1739, I.III.xiv).

It is a psychological peculiarity of our minds that we develop a tendency to associate a certain kind of event with events like the cause, and the expectation prompted by the cause leaves us with the impression of a necessary connection which we mistakenly read into nature. This explanation relies on the propensity of one belief to give rise to another, and clearly presupposes the notion of causation at issue. If causation of this sort were available to us for inspection, it would undermine Hume's initial critique. But as Stroud says, "Hume does not think that we actually perceive the necessity of the connection between any two events, even events that occur in our minds. If we did, then we could get the idea of necessity directly from one of our internal experiences" (Stroud, 1977, p. 84). Our impression of necessity is just a feeling that arises in the mind when one mental event causes another; it is not an impression of a causal connection between two events. The analysis which Hume provides in his answer to the second of his questions would be redundant if this were not the case.

A concise summary of his answer to the second question is given by the following definition:

> We may define a cause to be "An object precedent and contiguous to another, and where all objects resembling the former are placed in like relations of precedency and contiguity to those objects that resemble the latter" (Treatise 1739, I.III.xiv).

The nub of Hume's analysis is what he calls the "constant conjunction" of events which fall into kinds of mutually resembling objects. Objects of each kind are paired off with one another in virtue of their *conjunction* – a term which is to be understood as the astronomer, rather than the logician, uses it. The modern logician wants to reserve the term "conjunction" for a sentential connective, and would use some other expression, say "occurring together", for the relation between objects involved in causation. Now, without some such relation of occurring together, the mere resemblance of events would leave us with the unhelpful statement that events of one kind occur and events of another kind occur. But Hume well understood that his notion of constant conjunction involves some "relation among objects" pairing off events:

> The idea, then, of causation must be deriv'd from some *relation* among objects; and that relation we must now endeavour to discover. I find in the first place, that whatever objects are consider'd as causes and effects, are *contiguous*; and that nothing can operate in a time or place, which is ever so little remov'd from those of its existence. Tho' distant objects may sometimes seem productive of each other, they are commonly found upon examination to be link'd by a chain of causes, which are contiguous among themselves, and to the distant objects; and when in any particular instance we cannot discover this connexion, we still presume it to exist. We may therefore consider the relation of CONTIGUITY as essential to that of causation; at least may suppose it as such, according to the general opinion, till we can find a more [fn. referring to the later section I.IV.v] proper occasion to clear up this matter, by examining what objects are or are not susceptible of juxtaposition and conjunction. (Treatise 1739, I.IV.ii)

A distinction is drawn here between mediate and immediate causation, and the contiguity requirement applies to the latter. Distant causation is allowed on the condition that there is a chain of events, in which each is an immediate cause of the next, linking distant cause and effect. A more general notion of mediate causation can thus be defined once a concept of immediate causation involving the contiguity constraint is available, and so the latter remains the basic notion which Hume is concerned to delimit in his original definition.

The contiguity requirement can be motivated with reference to examples like the famous one of Russell's in which a whistle blows every weekday in the afternoon in a certain London factory and workers pour out of a Manchester factory a few minutes later. It would be absurd to suggest that a whistle they couldn't even hear caused the Manchester workers to down tools for the day, and there is no suggestion of an intermediate chain of events mediating a link between London and Manchester. Causes are accordingly restricted to events near their effects. Roughly, the general idea is that there are regularities between events of various kinds which we wouldn't want to say are causally connected, and the contiguity requirement is called upon to eliminate such coincidences. Note that the same argument would apply against a whistle's blowing yesterday in Manchester as the cause of the workers leaving the Manchester factory today, so the contiguity includes both spatial and temporal nearness.

Whatever might be said in support of the reasonableness of appealing to contiguity, however, it shouldn't be forgotten that the constant conjunction account *requires* some analysis of the conjunction, or occurring together, relation. Given a relation of occurring together defined between particular events, a relation of constant conjunction can be defined between kinds of events X and Y by saying that each event of kind X stands in a relation of occurring together with an event of kind Y. How is this relation of occurring together to be defined? If the analysis is not to be circular, or uninformative by reason of relying on a concept Hume regards as synonymous with causation, it cannot be explained in terms of causal connection, but must be explicitly defined in non-synonymous terms. Hume thought a definition satisfying these requirements was available in terms of the spatial and temporal contiguity of events, which can be motivated along the lines indicated above. Should this motivation be found wanting, some other account of the "occurring together" relation must be provided unless the whole framework of the analysis is to be radically rethought.

Hume did express doubts about the contiguity condition. In section I.IV.v of the *Treatise* he considers that "Thought ... and extension are qualities wholly incompatible", and worries about "the soul['s] ... *local conjunction* with matter", which leads him to wonder whether "it may not be improper to consider in general what objects are, or are not susceptible of a local conjunction" (pp. 234–235). The absurdities of "endeavouring to bestow a place on what is utterly incapable of it" (p. 238) may have convinced Hume that "our perceptions are not susceptible of a local union". But he is not then at liberty to conclude that "as the constant conjunction of objects constitutes the very essence of cause and effect, matter and motion may often be regarded as the causes of thought, as far as we have any notion of that relation" (p. 250). For the *relation* of constant conjunction depends upon a relation of occurring together, and the only interpretation of this Hume has offered is one in terms of contiguity.

A further element is required, for as it stands, this analysis gives no account of the asymmetry of the causal relation, i.e. it makes no distinction between cause and effect, since the contiguity relations are symmetric. Asymmetry is introduced with the requirement that the cause precedes the effect in time. Putting these pieces together, then, the regularity theory defines the causal relation as follows:

x causes y if and only if x is of some kind X and y of some kind Y which are

(a) constantly conjoined and y is the Y-event occurring together with x (an X-event), and

(b) the X events uniformly precede the Y events with which they occur together.

This definition provides Hume with an answer to his second question of what distinguishes causally related events without the need to appeal to any notion of causal power, efficacy or necessity. No features are mistakenly attributed to nature which are in reality only features of our psychology.

3. Some Difficulties with Hume's Analysis

Objections have been raised against the regularity theory, as Hume's theory is often called, ranging from technical details concerned with including and excluding exactly the right cases to general matters of principle. Sometimes technical problems raise matters of principle. Not all regularities, as we have seen, provide cases of causally connected events, and the question arises whether the spatio-temporal contiguity constraint suffices to preclude such counterexamples. Thomas Read pointed out that night invariably follows day.[2] This is perhaps best seen not so much as a counterexample to the sufficiency of Hume's condition as a reminder that the definition specifies causation as a relation between events with clearly defined spatial and temporal boundaries – objects "susceptible of juxtaposition". "Night" and "day" don't refer to any such thing, but merely indicate general conditions, and whatever kind of regularity they give rise to, it isn't a constant conjunction as defined and so not a case in which the *definiens* is satisfied but not the *definiendum*. Responding to Read's example in this way emphasises an ontological presupposition of Hume's own version of his regularity theory, namely the existence of entities of a certain kind with definite spatio-temporal boundaries.

A second challenge to the sufficiency of the condition is provided by examples in which several distinct events are regularly produced by a single common cause. Symptoms of diseases are often of this kind, for

[2] "But no reasoning is more fallacious than this – that, because two things are always conjoined, therefore one must be the cause of the other. Day and night have been joined in a constant succession since the beginning of the world; but who is so foolish as to conclude from this that day is the cause of night, or night the cause of the following day?" (Read, 1785, Ch. IV, p. 253).

example the appearance of swellings under the cheeks followed by severe headache and rise in temperature which are all symptoms of mumps. Again, a fall in pressure causes a swing in the barometer needle and a subsequent storm. Reichenbach (1928) discussed the problem of distinguishing causally connected processes from what he called "unreal sequences" in which events in a spatio-temporally continuous series are not causally linked such that earlier ones cause the immediately succeeding ones; for example, when the beam from a torch (or rather a laser) is swung so that the image falling on a distant object moves across the surface at a speed greater than that of light. The trajectory of the image is continuous, but not causally linked. The common effects are constantly conjoined with one preceding the other, and so satisfy the two conditions in Hume's definition; but the earlier one is not the cause of the later.

Reichenbach's solution is to repeat the process with a slight variation in the cause – more specifically, in the kind of event featuring as cause. The variation should not be so much as to constitute an essentially different kind of event, but just sufficient to "mark" events as variations of the putative cause kind. If it really is the cause kind that is thus marked, a concomitant variation will be noticed in the effect kind, and the mark may be said to be transmitted from cause to effect. No such transmission is observed if the event kind marked corresponds to one of two effects of a common cause.

Another, albeit less illuminating, approach is to adopt a similar manoeuvre to that involved in extending the basic concept of immediate causation to cover mediate causation. There the strategy was to extend the original definition by adding a disjunct, thus

> x causes y if and only if *either* x is the immediate cause of y *or* there are events $z_1, \ldots, z_n$ such that x is the immediate cause of z_1 and z_1 is the immediate cause of z_2 and ... and z_n is the immediate cause of y.

In the present case, however, the point of the added clause is to revise the concept just defined by restricting rather than extending it, giving us yet a new concept conveniently called strict causation:

> x strictly causes y if and only if x causes y *and* there is no event z such that z causes x and z causes y.

An objection to the effect that the Humean definition doesn't provide a necessary condition for the causal connection of events might be motivated along the following lines. A paradigm case of causation is the switching on of the light in a room by pressing the switch on the wall. Now we are all familiar with the situation where, on a specific occasion, the switch is pressed but the light doesn't come on. The explanation might be simple – perhaps the tungsten filament in the light bulb has burnt out. Nevertheless, the condition of the regularity theory is broken; there is no constant conjunction between turnings of the switch and lightings of the light. Two lines of reply are available to the regularity theorist. First, he might say that the objection builds on an unwarrantably strict interpretation of the theory. When we say that such-and-such causes so-and-so, it is understood that other factors are operative in addition to those explicitly mentioned as the cause. In the case of the pressing of the switch, these would include the assumption that the mechanics of the switch are in order, that the wiring is in reasonable condition, that the bulb is in working order, etc., where explicit descriptions can be provided of what "being in order" presupposes in each case. In other words, the cause as actually described is not itself sufficient for the effect, but is rather a non-redundant part of a set of relevant conditions which are jointly sufficient for the effect.

This first line of reply is not satisfactory as it stands, however. By Hume's own argument, the notion of sufficiency at issue in causation is not a logical or analytic one. It involves all the relevant factors being present, the absence of one or more rendering the cause insufficient. But of all the states of affairs obtaining in the world, which are to be counted as relevant and which not? If Hume's argument against any special notion of necessary connection holds water, then by parity of reasoning it also applies to the circumstance of being relevant. The fact that one state of affairs is relevant to the explicitly mentioned feature as a part of the total cause is just as unobservable as the original necessary connection. It looks like this first line of reply invokes a notion of relevance which, if not synonymous with necessary connection, is of much the same kind, and is at odds with the spirit of the regularity analysis. This point was further emphasised by Nelson Goodman in his discussion of counterfactual conditionals (Goodman, 1947).

This first line of reply errs, it might be thought, in accepting too much of the presuppositions of the objection. A second line of reply insists on the relational analysis of causation, the terms "cause" and "effect" referring to definite entities called events which exist in space and time even if they don't persist over time as do ordinary physical objects like tables and chairs. Now there is no question of an ordinary description of a persisting physical object – say "that man standing in front of the shop window" – mentioning all, or even an appreciable number, of the properties actually possessed by the object in question. Similarly, the fact that the actual description of the event said to cause another in a singular causal statement only mentions some feature of the event doesn't detract from the fact that the event actually has innumerable other features, some of which are necessary for the bringing about of the effect. Davidson – a well-known proponent of events in the sense required for this second line of reply – answers Mill's objection that a fall cannot be the cause of death because the circumstance of his weight must be included in the cause with the comment that

> ...if it was Smith's fall that killed him, and Smith weighed twelve stone, then Smith's fall was the fall of a man who weighed twelve stone, whether or not we know it or mention it. (Davidson, 1967, p. 150)

This second line of reply emphasises the difference between taking the notion of an event seriously and merely using it as a *façon de parler* without commitment to the existence of entities it ostensibly refers to. Probably Hume wasn't himself so sure on this matter. For if he was, he wouldn't have begun his critique of the concept of causal necessity by first insisting that no logical connection could be involved between cause and effect. Davidson (1963) points out that logical connections are sustained between descriptions of things, and it is perfectly possible on the relational view of causation that the description of the cause entails the existence of the effect, for example in "The cause of fire caused the fire". Identity, on the other hand, is a relation sustained by the entities themselves, and Davidson emphasises that the cause is always distinct from (i.e. non-identical with) the effect, even when a description of the one entails the existence of the other.

Objections might still be raised. Kant suggested that the causing of a depression in a pillow by a heavy object provides an example of simultaneous causation. A similar example is the rotation of a ball on a string causing a tension in and elongation of the string. Statics provides examples where no motion is involved: a pillar standing in the appropriate position holds up a bridge; a person leaning against a spring-door holds it open; and so on.[3] If these counterexamples are allowed and temporal precedence abandoned, another criterion of causal priority – what distinguishes cause from effect and renders the causal relation asymmetric – is needed. Reichenbach (1928) thought this was needed in any case because he wanted to carry through Leibniz's idea of reducing temporal order to causal order. Leibniz (1715, pp. 201–202) doesn't seem to realise that if "earlier than" is to be defined in terms of "causes" without circularity, the asymmetry of the latter must be given an independent explanation. The solution Reichenbach proposed

[3]Irreversible thermodynamics provides further examples. "For certain systems the fluxes at a given instant depend only on the values of the affinities at that instant. Such systems are referred to as 'purely resistive'. ...a very large fraction of the systems of interest ... are purely resistive." (Callen, 1985, p. 312).

– the same as his solution to the common cause problem – has not gone without criticism; but some philosophers believe that he was right in thinking that causal priority should be accounted for independently of temporal precedence.

The examples from statics might be construed as counterexamples of a kind which are, *prima facie*, singular causal statements, but which are difficult to cast in the relational mould. Leaning against a door is a state a person is in rather than an event in which he participates, and it seems far-fetched to represent the proposed causal situation as a relation between two events. Opinions may differ on the appropriate analysis of putative counterexamples of this kind; but one source of examples is particularly embarrassing for the relational theorist. Consider

(1) The fact that the earth and moon have such and such masses and stand at such and such a distance causes it to be the case that they revolve around a common centre of gravity,

which is one way of expressing Newton's thesis that gravitational force causes the moon's orbital motion. Newton speculated on an abundance of forces in nature by which objects attract and repel one another, and interact chemically, although he only succeeded in formulating the gravitational force. As already mentioned, the intelligibility of gravitation was hotly contended on the grounds that it involved action at a distance. Nevertheless, the law survived the imputations of the Cartesians as the only serious contender in the field, and determined the pattern of development of natural science from the eighteenth century on. More than two hundred years were to elapse before the first coherent alternative was formulated; and even then, Newtonian theory retains a position other scientific theories can hardly rival in their own spheres of application. Yet here we have the spectacle of a philosophical theory of causation, supposedly built upon the foundations of observation and recognising no non-analytic *a priori* truths, which doesn't seem able to accommodate gravitational attraction. Even if (1) could somehow be cajoled into relational form by some appropriate construal of events (we haven't yet been provided with any clear criteria for identifying and distinguishing events beyond Hume's directive that they be "susceptible of juxtaposition"), the analysis precludes cause and effect being simultaneous. But Newton's law requires that the orbital motion caused by the gravitational force occur when the force operates; if mass were miraculously divested of its attractive power, the moon would instantaneously fly off its orbit at a tangent. Surely no adequate *concept* of causation can reasonably conflict with scientific theory of any sort, let alone one of the most successful theories of them all.

Newton's contemporary Leibniz opposed the idea of action at a distance with another proposal, the apparent action being in reality the occurrence of simultaneous events in distant bodies standing in preestablished harmony. This doesn't call upon action by contact to render the effects intelligible, and as Duhem (Duhem, 1893, pp. 125ff.) points out, scientists after Leibniz generally agreed that the action of one body on another remains as much in need of explanation when bodies are in contact as when they are not, even if they saw little point in denying the reality of the action in favour of preestablished harmony. More modern views treat the contact of two bodies in collision as a macroscopic description which is not preserved at the microscopic level, where there is no question of massive bodies literally touching one another.

The requirement of contiguity has not always been preserved by empiricist-minded philosophers who have sought to retain what they regard as the essential insight of Hume's regularity account, namely the avoidance of modal concepts in the analysis of the concept of a general law, without the specific ontological commitment to events. The following section describes the analysis developed by Hans Reichenbach towards the end of his life. It is representative of the kind of extensional analysis typical of the logical positivists.

4. Reichenbach's Analysis of Laws

Reichenbach outlined a fairly sophisticated view of lawlikeness in *Elements of Symbolic Logic* (1947) which, after a certain amount of criticism, he adjusted and developed in considerably more detail in his posthumously published *Nomological Statements and Admissible Operations* (1954). The latter book received an unfavourable review on publication by Hempel (1955). But in a general review of the problem of law, Jobe (1967) distinguished Reichenbach's theory as by far the best then available. Reichenbach lays down a number of formal conditions for lawlikeness. A simplified presentation the sort of thing involved is given in the next section. But he recognised that formal conditions are not sufficient, and elaborated a further notion of verifiability, which is discussed first in this section.

The first requirement Reichenbach lays down is a non-formal one, that a fundamental law be *verifiably true*, which he explains as meaning "it is verified as practically true at some time during the past, present or future history of mankind" (Reichenbach, 1954, Def. 1, p. 18). Reichenbach was clearly concerned to eliminate the modal character of an undefined dispositional term "verifi*able*" by explaining it in terms of being confirmed at some time – past, present or future. But these words have led to some unfortunate misunderstanding of Reichenbach's intentions – an unhappy fate for the first major concept of his book.

The emphasis on confirmation led several critics to interpret him as saying that a law is merely confirmed to a high degree *rather than* being true. So Hempel read him in his review, and Carnap followed suit:

> My friends argued that they would prefer to say, instead of "true", "confirmed to a high degree". Reichenbach, in his book *Nomological Statements and Admissible Operations* ... comes to the same conclusion, although in different terminology. By "true" he means "well established" or "highly confirmed" on the basis of available evidence at some time in the past, present or future. But this is not, I suspect, what scientists *mean* when they speak of a basic law of nature. By "basic law", they mean something that holds in nature regardless of whether any human being is aware of it. (Carnap, 1966, p. 213)

Although there is an important point in the last sentence of this passage, the initial part interprets Reichenbach incorrectly. Reichenbach actually says in his introduction that

> Being laws of nature, nomological statements, of course, must be true; they must even be verifiably true, which is a stronger requirement than truth alone. (Reichenbach, 1954, p. 11)

It couldn't be more clearly put that truth is part of what is involved in a law of nature. The point is driven firmly home by Jobe, 1967, pp. 374–375, and to avoid any misunderstanding the expression "verifiable and true" is used here instead of Reichenbach's "verifiably true".

Carnap sought to characterise laws by trying to distinguish conditions on logical form necessary for laws of nature, and counting as basic laws those statements which fulfil the conditions on logical form and are true. Lawlikeness, in other words, is conceived as a purely formal matter. "The problem of defining 'basic law' has nothing to do with the degree to which a law has been confirmed ... the problem is only concerned with the meaning that is intended when the concept is used in discourse by scientists" (Carnap, 1966, p. 213). Reichenbach's approach is fundamentally different from this. Taken in isolation, formal conditions are not sufficient to guarantee "generality in a reasonable sense", which Reichenbach thought must be broad enough to include a notion of inductive extension. For accidental generalisations

> may obtain even if no reference to individual space-time regions is made; for instance, the statement "all gold cubes are smaller than one cubic mile", may possibly be true. (Reichenbach, 1954, p. 11)

Thus, "All the coins in my pocket are silver" is not universal in Reichenbach's sense of containing no individual term "defined with reference to a certain space-time region, or which can be so defined without change of meaning" (Reichenbach, 1954, Def. 24, p. 32), and so disqualified as a law. But "All gold cubes are smaller than one cubic mile" is universal, and satisfies his other formal conditions too, yet is not a law. What it lacks, in Reichenbach's view, is inductive generality:

> when we reject a statement of this kind as not expressing a law of nature, we mean to say that observable facts do not require any such statement for their interpretation and thus do not confer any truth, or any degree of probability, on it. If they did, if we had good inductive evidence for the statement, we would be willing to accept it. For instance, "all signals are slower than or equally fast as light signals", is accepted as a law of nature because observable facts confer a high probability upon it. It is inductive verification, not mere truth, which makes an all-statement a law of nature. In fact, if we could prove that gold cubes of giant size would condense under gravitational pressure into a sun-like ball whose atoms were all disintegrated, we would be willing also to accept the statement about gold cubes among the laws of nature. (Reichenbach, 1954, pp. 11-12)

What, then, does Reichenbach put into this notion of verifiability which he thinks so important? A high probability, for one thing; but that is not sufficient. Reichenbach formulates other conditions on the confirmation of an all-statement $\forall x(Px \supset Qx)$ in terms of the conditional probability of something's being Q given it is P. The formal conditions preclude vacuous antecedents, and the exclusion of reference to specific spatio-temporal regions precludes verification by only a small number of P's. The class of P's must be *open* in the sense that there are many more of them than those which happen to have been examined for the property Q.

Reichenbach requires that not only the conditional probability of something's being Q given it is P must be high. The conditional probability of something's being $\sim P$ given that it is $\sim Q$, $p(\sim P/\sim Q)$, must also be high. The one doesn't necessarily follow from the other, and so the requirement is a substantive one. Reichenbach shows that the deviation, d, from 1 of $p(Q/P)$ (i.e., $p(Q/P) = 1 - d$) is related to the deviation, d', of $p(\sim P/\sim Q)$ from 1 by

$$\frac{d}{d'} = \frac{1 - p(Q)}{p(P)}$$

For example, the probability of "Houses are red" is not high, although the probability that something not red is not a house is high. Since the probability of not being red, $1 - p(Q)$, is high, and that of being a house, $p(P)$, is low, the ratio d/d' is high. A high probability of the conditional probability directly related to the contraposition of a hypothesis is therefore compatible with a low conditional probability directly related to the hypothesis originally formulated. This, Reichenbach claims, supplies an answer to Hempel's paradox of confirmation (Hempel, 1965), explaining why what confirms "All non-Qs are non-Ps" doesn't confirm "All Ps are Qs".

It also serves, he claims, to block "All gold cubes are smaller than one cubic mile", which doesn't satisfy the condition because there is very little direct evidence to support the contrapositive "Anything at least one cubic mile in volume is not gold" – i.e. there are no grounds for assigning $p(\sim$ gold / $\sim$ less than one cubic mile in volume) a high value. It has a low probability because we can quite easily imagine such a large piece of gold; no facts seem to count against this possibility. Compare "All objects made by man are under 400 meters high", which can't be assigned a high probability because it might well not be true for all time. On the other hand, the probability of an object over 400 meters high not being man-made is high – the observation of high mountains, for example, provides strong support – but this doesn't influence the low estimate of the probability of the original statement, which is therefore not lawlike either. It is a moot point, however, whether the interpretation of probability here is consistent with the general restriction to an extensionalist account.

Finally, a high value of $p(Q/P)$ provides no guarantee that there are no exceptions to "All Ps are Qs". "It would be too strong a condition to require that *there be* no exceptions", Reichenbach says. "In some sense, there exists general evidence that exceptions will occur, because too many laws of physics have later turned out to be merely approximately true. But we must be unable to describe conditions upon which an exception ... could be expected. In other words, there should be no *specific evidence* that the general implication considered is subject to exceptions" (Reichenbach, 1954, p. 132). Accordingly, a condition to the effect that there is no evidence of a property C such that the degree of confirmation of $\forall x((Px \wedge Cx) \supset Qx)$ is less than that of $\forall x(Px \supset Qx)$ is imposed. There should, in fact, be evidence that there is no such property C. Thus, the probability that swans are white is high, but some swans in Australia were found to be black. The probability that something is white given that it is an Australian swan is less that it would be given that it is simply a swan. Despite the fact that $\forall x(Px \supset Qx)$ logically implies $\forall x((Px \wedge Cx) \supset Qx)$ might be less than $p(Q/P)$ and independent evidence must be considered to establish just what is the case. (Intuitively, it might be thought that whenever A confirms B and B implies C, then A confirms C since C would seem to say no more than anything implying it. But this principle, though not without some *prima facie* plausibility, is evidently denied by Reichenbach.)

5. Reichenbach's Formal Conditions on Lawlikeness

The notion of verifiability complements a set of purely formal conditions on lawlikeness in Reichenbach's account. Here his strategy is first to define a set of *original nomological statements*, and then to define the *derived nomological statements* and finally, *relative nomological statements*. Statements of the first kind are intended to be the fundamental laws, including the laws of logic. Derived nomological statements are, broadly speaking, logical consequences of the set of original nomological statements, and relative nomological statements are consequences of original or derived nomological statements together with particular factual information. The law that the period of a pendulum is given by $2\pi\sqrt{l/g}$, where l is the pendulum's length and g the acceleration due to gravity, for example, is derived from laws together with factual information about which particular forces are acting.

In order to ensure "generality in a reasonable sense", original nomological statements satisfy a number of conditions. First and foremost, they must be generalisations – *all-statements* – which means that, when reduced to prenex normal form (so that all the quantifiers occur at the front of the formula), there is at least one universal quantifier. But this is not enough, and Reichenbach imposes further conditions. Fundamental laws must be universal, which means that an original nomological statement contains no individual term "defined with reference to a certain space-time region, or which can be so defined without change of meaning" (Reichenbach, 1954, Def. 24, p. 32). But there are several cases of statements that have been counted as laws which do make specific reference to specific bodies, and therefore to specific space-time regions – Galileo's law of free fall and Kepler's laws. Moreover, Goodman (1947) points out that it is always possible to express what these laws say in terms of sentences built up from predicates whose syntactic form gives no indication of reference to specific bodies. Instead of talking about bodies on the surface of the earth, for example, we could talk of terrestrial bodies.[4] Furthermore, general statements which don't make reference to objects can be equivalently expressed in ways that do; for example, by conjoining a tautology "John's hair is blond or it is not".

Regarding the first objection, laws such as Kepler's which deal with planets in the solar system are not nomological statements according to

[4]Ramsey made this point in notes written in 1928 (but first published in 1978). "If we put in enough detail", he says, "we shall (unless the world repeats itself endlessly with just a few details different each time) get a true generalization which mentions no particular portion of space-time but this would not be a law of nature" (Ramsey, 1978, pp. 130–131).

Reichenbach, but only relative nomological statements. A statement is relative to some matter of fact p if it is deductively derivable from p and a nomological statement, s. Kepler's laws are derivable from Newton's laws, which are universal, together with certain information about the relative masses and velocities of the planets and the sun. Against this, Nagel (1961) argues that if non-lawlike premises are allowed to figure in the derivation of laws from others, then we must face the consequence that a statement such as "All the screws in Smith's car are rusty" is lawlike, since it is presumably a law that iron screws rust when exposed to oxygen, etc., and this conjoined with the additional premises that all the screws in Smith's car are iron and have been exposed to oxygen, etc., implies the unwanted conclusion. But Jobe counters that logically consistency appears to be on the side of the explication. For a law is not the sort of thing that can be falsified by an accident, and yet

> ... a mere accident in the form of, say, a near approach or collision with an interstellar body could impart to one or more of the planets velocities such that, contrary to Kepler's first law, they would pursue hyperbolic rather than elliptical orbits with respect to the sun. (Jobe, 1967, p. 380)

Nevertheless, Kepler's laws are nomological statements relative to conditions of great stability and permanence, at any rate on a human time scale, whereas "All the screws in Smith's car are rusty" is nomological relative to quite ephemeral circumstances. This is quite sufficient to mark a difference between the two cases. A similar tale can be told about Galileo's law, which depends upon the earth's being a sphere so that all its mass can be regarded as being concentrated at its centre. If the rate of rotation of the earth were greater than it is, the deviation from a sphere would be greater than its present negligibly small deformity.

On the second point, Reichenbach is aware that reference to specific bodies or regions must be interpreted as an essential feature of accidental generalisation but an inessential feature of laws. Accordingly, the test of lawlikeness must be whether there is an appropriate formulation which does not make specific reference to particular bodies or regions, rather than that all equivalent formulations lack such reference. He therefore defines a synthetic statement as universal if "it cannot be written in a reduced form which contains an individual term" (Reichenbach, 1954, Def. 25, p. 33). The notion of a reduced form is explained in a series of definitions, the effect of which is to eliminate redundant parts and to contract a statement to a shorter equivalent form. Thus, "Peter weighs as much as Paul" is discounted as a fundamental law, although it is equivalent to the general statement "Everyone weighs as much as Peter if and only if he weighs as much as Paul". But Reichenbach recognises that other counterexamples are forthcoming unless laws are defined only for a language which, as he puts it, can reasonably be regarded as scientific. The following passage illustrates the sort of restriction Reichenbach had in mind.

> A certain ambiguity arises because a natural language is often capable of different rational reconstructions. The term "polar bear", for instance, can be interpreted as meaning a bear living in the polar regions of the Earth, in which interpretation it would be an individual-term. It could also be defined as a biological species with certain general characteristics, for instance, as a bear with a white skin, etc. In such cases, we have two rational reconstructions which are not logically equivalent, though perhaps practically equivalent. As a consequence, a statement which in one rational reconstruction is nomological, may not be so in another reconstruction of conversational language. This ambiguity, however, offers no difficulties, since the class of nomological statements is defined only for a certain reconstruction of language. If a statement of conversational language is given, it would be meaningless to ask: is this statement really nomological? There is no such thing as an absolute meaning of the terms of a natural language. A classification of statements such as expressed in categories like "analytic" or "nomological" refers to a given rational reconstruction of language. Whether this reconstruction is adequate, is to be investigated separately. (Reichenbach, 1954, p. 34)

A further condition Reichenbach imposes is that of *exhaustiveness*. This is intended to rule out vacuously true statements such as "All unicorns are pink", which have antecedents not true of anything, but also other types of vacuousness. For example, consider

(2) $\forall x \exists y(Fxy \supset Txy)$

where Fxy means "x is the father of y" and Txy means "x is taller than y". Fxy is not always false (for every value of x and y), but it is true that

(3) $\forall x \exists y {\sim} Fxy$

and so (2) is vacuously true.

The requirement works along the following lines. The quantifiers are assumed to have been moved to the front of the formula. A statement's being *exhaustive in major terms* is defined first, for which purpose the main binary truth functional operator is expanded in disjunctive normal form. $\forall x(Ux \supset Px)$, for example, ("All unicorns are pink"), which already has its quantifiers at the front, has a major truth functional operator "$\supset$" and can be expanded thus:

(4) $\forall x((Ux \wedge Px) \vee ({\sim}Ux \wedge Px) \vee ({\sim}Ux \wedge {\sim}Px))$.

A *residual in major terms* is then defined as any statement obtained from the expanded form (i.e. (4) in the case of the present example) by removing one or more of the disjuncts. A statement is then said to be *exhaustive in its major terms* if none of its residuals in major terms is true and verifiable. The residual

$$(5) \qquad \forall x((\sim Ux \wedge Px) \vee (\sim Ux \wedge \sim Px))$$

of (3), for example, is true and verifiable since there are no unicorns, and therefore $\forall x(Ux \supset Px)$ is not a law. A further notion of being *exhaustive in elementary terms* is then defined along the same lines, except that the reduction to disjunctive normal form is carried through for all connectives and not just the principal binary operator. Finally, a statement is said to be *exhaustive* if it is exhaustive in both major and minor terms. The effect of this requirement is that a law cannot say too little; if a stronger statement can be made, then the weaker statement obtained by adding an extra, vacuous disjunct cannot be called a law.

Against this procedure it might be objected that some of the most fundamental laws of nature are in fact vacuous, dealing with ideals of natural order rather than the real thing. The Charles-Boyle, or ideal, gas law $PV = nRT$, for example, is appropriate for ideal gases which real gases approach under low pressure and high temperatures. Raoult's law, stating that the vapour pressure of a solvent over a solution is reduced from the vapour pressure of the pure solvent in proportion to the mole fraction of solvent in solution, deals with ideal solutions, which like ideal gases, may be approached in certain limiting cases. The Hardy-Weinberg law states that the gene ratios in a population remain constant over the generations in the absence of any influences favouring the selection of particular genetic characteristics. But Darwin's principle of natural selection denies that populations remain undisturbed. Newton's first law of motion concerns bodies not acted upon by any external forces; but given his law of gravitation, there are no such bodies. Laws such as these seem to be about ideal entities rather than the real objects over which the quantifiers in Reichenbach's explication presumably range. Doesn't the exhaustiveness requirement make the extensional analysis too idealistic an account of scientific law?

It has been maintained that appeal to picturesque idealisations is not essential, however, and the function of these laws is certainly in the explanation and prediction of the behaviour of real objects. Newton's first law, for example, can be expressed in the form "For any body x, if no resultant force acts on x, then it is at rest or in uniform motion (relative to the fixed stars)". A body is subject to a resultant force if all the forces acting on it are such that they don't cancel one another completely and there is a net overall force in a certain direction. Arthur Pap (1958) objected to this reformulation that the law is actually used in its vacuous formulation, for example when calculating the tangential velocity of a body moving under the influence of a central force. The tangential velocity at a given instant is the velocity the body would have if the central force were removed at this instant and the body continued in a straight line with its inertial movement. This counterfactual would not follow, Pap maintains, from the suggested reformulation, but only from the vacuous statement of the law. But, as Jobe (Jobe, 1967, p. 379) points out, this argument is based on a fallacy that being acted on by no force, as the antecedent of the counterfactual states, is being acted on by no resultant force. Perhaps Pap thought "acted on by no resultant force" defines a subset of the bodies acted on by no forces, rather than vice verse, because "acted on by a resultant force" defines a subset of the bodies acted on by a force. However that may be, it seems that vacuous formulations can be dispensed with and the exhaustiveness requirement upheld. (In more recent times, Nancy Cartwright (1983) has presented a different case for the vacuousness of fundamental laws.)

Reichenbach goes on to extend the concept of exhaustiveness to deal with another problem, closely connected with the problem of individual terms and restricted spatio-temporal regions which led him to the universality requirement. The problem is best illustrated with an example. Helmholtz was the first man to have seen a living human retina, and this property, unique to Helmholtz, can be used to form a definite description of him. But a sentence such as

For all x, if x is a man that has seen a living retina, and no other person has seen a living retina before x, then x contributed to the establishment of the principle of the conservation of energy,

which contains a tacit definite description of Helmholtz, is not universal according to Reichenbach's definition because it can be reduced to a form which makes the individual term explicit as a definite description. By using some, but not all, of the information employed in the description, however, the basic problem of the occurrence of a disguised reference to a restricted spatio-temporal region is still with us in the form of a statement no longer equivalent to one in which an individual term occurs. For example, "All stars seen by any man who saw a living human retina before any other man were at least of the 11th magnitude" expresses no more than a technological limitation on the telescopes available at the time of Helmholtz and can hardly be regarded as a law of nature. But the statement does not imply the existence of such a man. It therefore cannot be equivalently transformed into a statement with the definite description operator and is, accordingly, universal. Another restriction is necessary to eliminate such cases.

To deal with his problem Reichenbach requires that original nomological statements be *unrestrictedly exhaustive*, a notion defined in terms of properties Rx specifying that x occupies some restricted spatio-temporal region, and the basic notion of exhaustiveness already defined. The spatio-temporal predicates are introduced in the following way. Suppose Lx stands for "x saw a living human retina before any other man" and Rx for "x exists in such and such a restricted spatio-temporal region". The statement

(6) $\forall x(Lx \supset Rx)$

is true and can be verified since the only person to do L satisfies R. In view of (6), then, any generalisation which contains the predicate L should not count as an original nomological statement. An example of a generalisation which does contain L is

(7) $\forall x \forall y((Ty \wedge Sxy \wedge Lx) \supset My)$.

where Tx means "x is a star", Sxy means "x sees y" and My "y is of at least the 11th order of magnitude". This statement is now expanded according to the procedure already outlined for exhaustiveness in major terms, with the difference that an additional disjunct Rx is added, thus:

(8) $\forall x \forall y(Rx \vee$
$(\texttt{Ant}(x,y) \wedge My) \vee (\sim\texttt{Ant}(x,y) \wedge My) \vee (\sim\texttt{Ant}(x,y) \wedge \sim My))$,

where $\texttt{Ant}(x,y)$ abbreviates the antecedent of (7). (8), formed by adding Rx in this way, is called an R-expansion of (7). The residual obtained from (8) by removing Rx is clearly true and verifiable, since it is just a reformulation of (7), and (8) (i.e. the R-expansion of (7)) is therefore not exhaustive. This means that the term "Rx" is redundant in (8) and therefore implicit in (7). The unreasonableness of the predicate L thus finds expression in the result that the R-expansion of (7) is not exhaustive. More generally, Reichenbach requires an original nomological statement to be unrestrictedly exhaustive, which means that there is no property R of being in a certain unrestricted space-time region such that, for any of the variables in the law, the R-expansion is not exhaustive.

The list of principal formal requirements on what is to count as a fundamental law or original nomological statement is completed with the condition that a fundamental law must be *general in self-contained factors*. The effect of this is to eliminate the possibility of simply adding to a statement which would otherwise count as a fundamental law a conjunct which is known to be true, such as "There are people".

Reichenbach went on to develop a concept of *admissible statements* upon which he builds a theory of counterfactuals, but the considerable detail involve won't be pursued here. We have seen something of the formal requirements Reichenbach imposed to exclude specific reference, whether explicit or implicit, and to preclude vacuous antecedents, as well as the distinction he draws between statements nomological relative to some particular matters of fact and fundamental nomological statements. Formal requirements are, however, at best only necessary conditions of lawlikeness. A sufficient condition for lawlikeness, in Reichenbach's view, calls upon verifiability together with the formal requirements.

Reichenbach's theory provided a natural starting point for Goodman's work on the problem of lawlikeness, which also emphasises the inadequacy of purely formal restrictions but points to severe difficulties with the notion of verifiability. Goodman concludes his comments on Reichenbach's initial attempt at a characterisation of laws by saying "the requirements Reichenbach sets up for nomological statements will be effective only if one places substantial and perhaps question-begging restrictions on the kinds of predicates that may be used" (Goodman, 1948, p. 414). We saw that Reichenbach intended his explication to apply only within a restricted language that can be "reasonably regarded as scientific", and the question arises whether some important problems have been swept under the carpet by assuming that a certain set of predicates is given in this way. Goodman certainly came to think so, and he developed his problem of the question-begging assumption that a range of appropriate predicates is given into his new riddle of induction in *Fact, Fiction and Forecast.*

6. On the Logical Character of Scientific Laws

The Achilles heel of any Humean analysis of scientific laws is the distinction between accidental and nomic universality. This is partly due to the general belief that extensional logic is the only valid logic available for representing the logical form of lawlike sentences. In 1948 Carl G. Hempel and Paul Oppenheim put forward their deductive-nomological model of explanation in which laws of nature play a crucial role.[5] For this purpose they adopted a rather strict notion of laws, according to which only true sentences can express laws. Since there are many lawlike sentences that have all the characteristics of a law except truth, however, every law can be represented by a lawlike sentence, but the converse does not hold. The logical structure of laws was thus distinguished from the empirical question of truth.

[5] All references are made to the reprint.

Hempel and Oppenheim's analysis seeks to provide individually necessary and jointly sufficient conditions for a sentence being lawlike and thus, by adding truth, to express a law: "Apart from being true, a law will have to satisfy a number of additional conditions. These can be studied independently of the factual requirement of truth, for they refer, as it were, to *all logically possible laws*, no matter whether factually true or false" (Hempel and Oppenheim, 1948, p. 265). According to Hempel and Oppenheim there are four such conditions which a sentence must meet to be lawlike: (1) it must have universal form, (2) it must be unlimited in scope, (3) it must make no reference to particular objects, and (4) it must contain only purely qualitative predicates.

The motivation behind these requirements seems to have broad intuitive appeal. First of all, lawlike sentences are statements of universal form, exemplified by 'All robins' eggs are greenish-blue' or 'All metals are conductors of electricity'. The first condition reflects the obvious idea that a law exists only if entities of a certain kind are all objects of such a law. If only some metals conducted electricity, we could not use this information to explain or predict anything about a particular piece of metal. But it is certainly not sufficient to characterise a lawlike sentence as having a universal form because some sentences are both true and of universal form but restricted to a certain place and a certain time. For instance, sentences like 'Every apple in this basket b at time t is red' have a limited scope by virtue of a reference to a specified object.

The second condition seems to yield a solution. A lawlike sentence must cover all objects in question, past, present, and future, in the universe. All robins' eggs are greenish-blue at all times, whereas every apple in a particular basket at a particular time is confined in space and time. Thus the scope of predication in a lawlike sentence must not be limited by any reference to a specific time or a specific place.

Further, a lawlike sentence should make no reference to particular objects. The idea of the third condition is that a law applies to an object irrespective of how we may conventionally name or classify this object. But there are many lawlike statements which we normally take to express laws of nature of which the third condition does not hold true . The first of Kepler's laws, "All planets move in elliptic orbits with the sun at one focus of the ellipse", for example. Hempel and Oppenheim attempt to cope with this problem by arguing that in contrast to sentences expressing mere accidental universality, Kepler's laws are known to be consequences of more comprehensive laws whose scope is unlimited and whose designators are not essential. Kepler's first law has a status of expressing a law of nature only if it is logically derivable from a fundamental law like Newton's laws. Thus, Hempel and Oppenheim follow Reichenbach in making a distinction between *fundamental* and *derivative* laws, and similarly between fundamental and derivative lawlike sentences. The latter is derivable from the former. The former, however, satisfy a certain condition of unrestricted scope and one of containing no uneliminable names for particular objects.

Nevertheless, problems remain. Nelson Goodman (1947) pointed out the need to impose certain restrictions upon predicates that are permissible in lawlike sentences. Consider the following sentence:

> Everything that is either a red apple in basket b at time t or a sample of ferric oxide is red.

Now, we may replace the predicate 'is either a red apple in basket b at time t or a sample of ferric oxide' with any arbitrary synonymous predicate such as 'is ferple'. This means that the above sentence can be expressed in the form

> Everything that is ferple is red.

This sentence is of universal form and contains no designation of particular objects, nor is it limited in scope. Yet, it is no more a fundamental lawlike sentence than the other sentence.

The solution, according to Hempel and Oppenheim, is to restrict predicates in a fundamental lawlike sentence to a purely qualitative ones, that is, ones where the explication of its meaning does not require reference to any particular object or spatio-temporal location. They mention terms like 'soft', 'green', 'warmer than', 'as long as', 'liquid', 'electrically charged', 'female', 'father of' as purely qualitative predicates, whereas 'taller than the Eiffel Tower, 'medieval', 'lunar', 'Arctic', and 'Ming' are not. They admit, however, that such terms suffer from a certain amount of vagueness since English as a natural language neither provides explicit definitions nor states unequivocally the meaning of its terms. The attempt to overcome the problem by introducing a formalised language will only help with respect to those predicates whose meanings are determined by definitions within the language. When it comes to the semantic interpretation of the primitive terms of a formal language there are no rigorous criteria for the distinction between permissible and nonpermissible interpretations. In spite of these difficulties of interpretation they conclude that there can be no doubt that a large number of purely qualitative predicates can be recognised to exist and that they are permissible in the formulation of fundamental lawlike sentences. Then Hempel and Oppenheim go on to give us a rigorous model theoretical characterisation of lawlike sentences based on the idea of purely qualitative predicates. But we will not go into the details here.

In his book *The Structure of Science* Ernst Nagel opposes the analysis proffered by Hempel and Oppenheim. He offers four kinds of considerations which seem relevant in classifying statements as representing laws of nature. These are (1) syntactical considerations in relation to the form of lawlike statements; (2) the logical relations of statements to other statements in a system of explanations; (3) the function assigned to lawlike statements in scientific inquiry; and (4) the cognitive attitudes manifested toward a statement because of the nature of available evidence. However, he doesn't claim that the conditions resulting from these considerations are sufficient or even, in some cases, necessary as a characterisation of a law of nature.

> Undoubtedly statements can be manufactured which satisfy these conditions but which would ordinarily not be called laws, just as statements sometimes called laws may be found which fail to satisfy one or more of these conditions. For reasons already stated, this is inevitable, for a precise explication of the meaning of "law of nature" which will be in agreement with every use of this vague expression is not possible. Nevertheless, statements satisfying these conditions appear to escape the objections raised by critics of a Humean analysis of nomic universality. (Nagel, 1961, p. 68)

What, then, are these conditions?

Nagel focuses on the question of the modal import of laws as consideration (1). He rejects the idea that nomic universality can be captured in terms of logical necessity or in terms of irreducible modal notions like "physical necessity". The first notion has the advantage that its meaning is transparent, but it faces grave difficulties since the formal denial of a law statement is demonstrably not self-contradictory, whereas the second notion is essentially obscure. He admits that there are contexts in which scientific laws are treated as if they are logically necessary and others where they are regarded as contingent. This is true of every sentence that can be associated with quite different meanings. But this does not tell us anything about the nature of scientific laws. Rather, it reflects the progress of science: "...the shifts in meaning to which sentences are subject as a consequence of advances in knowledge are an important feature in the development of comprehensive systems of explanation" (Nagel, 1961, p. 55). This assertion is reminiscent of the contextual view of laws. The difference is that Nagel thinks of it merely as a feature of our belief systems, whereas the contextualists, as we will see, take it to indicate something about the ontological relativism of scientific laws.

How, then, can the Humean, according to Nagel, distinguish between accidental and nomic universality? An answer to this query prompts consideration (2). He should impose a number of logical and epistemic requirements upon universal conditionals like having the form: for all x, if x is G, then x is H. The first, and most obvious, constraint is that it should be an *unrestricted* universal. This means that its scope of predication is not restricted to objects falling into fixed spatial regions or particular periods of time. Nagel rejects Hempel and Oppenheim's attempt to solve the problem by building on a semantic distinction between predicates that are "purely qualitative," whose meaning does not contain a reference to any particular object or space-time region, and predicates which are not purely qualitative. In addition, as we just saw, Hempel and Oppenheim introduced a further distinction between fundamental lawlike sentences, which contain only purely qualitative predicates, and derivative lawlike sentences. But Nagel finds this solution unsatisfactory partly because there was no fundamental lawlike statement from which Kepler's first law could be derived when he made his discovery, and partly because Kepler's first law is not derivable from fundamental laws alone. What is needed for such a derivation is "additional premises whose predicates are not purely qualitative" – premises which state the relative masses and the relative velocities of the planets and the sun.

According to Nagel, it is impossible on purely syntactic or semantic grounds to decide whether or not a universal conditional is unrestricted. The scope of predication may be finite, but the fact that it is finite cannot be inferred from the terms in the universal conditional that determine the scope of predication. The finite application must be established on the basis of independent empirical evidence. More important, however, is Nagel's recognition of a pragmatic identification of the scope of predication. He observes, as Hempel and Oppenheim did, that any restriction on the predication of attributes to an object could always be given a new name. His example is:

> For any x, if x is a screw in Smith's car during the time period a, then x is rusty during a.

A predicate "being a screw in Smith's car in a period of time a" may be replaced with the predicate like "being a scarscrew." Hence, we could reformulate the above conditional with an unrestricted universal:

> For any x, if x is a scarscrew, then x is rusty.

Nagel's point is that the syntactical structure of the new sentence does not reveal that the scope of predication is limited to objects satisfying a given condition during a finite period of time.

The unrestricted universality is only a necessary condition for a statement to express a law of nature. One would therefore expect that a candidate for being a law statement must satisfy further conditions. This search engenders consideration (3). If we look once again at Smith's car, we notice that it contains a finite number of rusty screws in a finite period of time. We are therefore in a position where we can hope to determine the truth-value of the conditional in question because we only have to examine a finite and fixed number of screws. Nagel thinks, furthermore, that if there is an indefinite number of screws in Smith's car, we may establish the truth of such an accidental universal conditional in two ways – either deductively from knowledge about all screws, for instance that they are made of iron and that they have been exposed to free oxygen, or inductively from the knowledge of a fair sample of screws in Smith's car. But this seems not to be the case concerning the evidence of unrestricted universal conditionals. Here, according to Nagel, it is a plausible requirement for an unrestricted universal statement to be called a law that the evidence for it is not known to coincide with the scope of predication and that its scope is not known to be closed by any further augmentation. This condition is important. It excludes an unrestricted universal conditional that is identical with a conjunction of statements expressing its total evidence from being a law. A law has the role of explaining and predicting. But it makes little sense to claim that a law explains or predicts a phenomenon if the law does not assert more than the evidence for it. If a law already contains a description of the phenomenon to be explained or predicted, it fails to explain or predict this phenomenon in any proper sense. What is also needed for calling a presumably true unrestricted universal statement a law is the assignment of a certain explanatory and predictive function to it. This rules out that the evidence on which a law is based is assumed to constitute the total scope of its predication.

In his last consideration (4) Nagel observes that laws can usually be recognised due to their special position in the corpus of our knowledge and due to the cognitive attitude we manifest toward them. He points out that the evidence by which we characterise a lawlike statement L as a law can be obtained directly or indirectly. The "direct" evidence consists of instances, whose properties fall within the scope of predication of L, whereas "indirect" evidence consists of instances that directly confirm other laws to which L is somehow inferentially connected. For instance, one sense of "indirect" evidence would be the case where direct evidence of Kepler's laws, Galileo's law, etc., serves as indirect evidence for Newton's laws. So a lawlike statement is called a law if both direct and indirect evidence support it.

Thus, according to Nagel, the Humean analysis of nomic universality should bring to light that a statement is often taken to express a law of nature because the statement occupies a distinctive position in the system of explanation in some area of knowledge, and because some evidence, satisfying certain specifications, supports the statement. Our attitude to such a universal statement is therefore more firmly settled than is our attitude to a universal statement without the required position and support of indirect evidence. We are not ready to abandon a universal conditional, which is considered to be a law, in the face of apparently contradictory evidence. This is due to the fact that laws are not only used as premises from which consequences are derived in accordance with the rules of formal logic but also function as rules of inference. When "a law is regarded as well-established and as occupying a firm position in the body of our knowledge, the law may itself come to be used as an empirical principle *in accordance with which* inferences are drawn" (Nagel, 1961, p. 67). His example is that the conclusion that a given piece of wire a is a good electric conductor can be derived from two premises, that a is copper and all copper is a good electric conductor. But the same conclusion can also be obtained from the single premise that a is copper if the principle of inference contains the rule that ' x is a good electrical conductor' can be derived from the statement ' x is copper'. This tendency concerning well-established laws may explain the view that lawlike statements express relations of logical necessity.

The Structure of Science is said to mark the end of positivism and logical empiricism. Nevertheless, as we have already noticed, in Nagel's analysis of laws there are certain elements which point in the direction of the contextual approach. Universal statements play different roles depending on the place these statements have in our system of knowledge and on the cognitive attitudes we have towards them. But the idea that the structure and organisation of our belief systems are important in distinguishing universal statements, which are accepted as laws, from universal statements that do not merit such a title, is something we also find in the Ramsey-Lewis account.

7. The Ramsey-Lewis Account

The problem of what is a law of nature and what form of sentences expresses such laws is not, as we've seen, merely a question of establishing syntactic and semantic criteria. The three criteria of being true, contingent, and a generalisation of the kind $\forall x(Px \supset Qx)$ are certainly not sufficient for being a law. Theory change of any kind may lead scientists to consider what were formerly accepted as laws to be mere generalisations, and vice versa, even though the sentences in question satisfy the syntactic and semantic criteria.

David Lewis (1973) develops a conception of laws of nature which copes with these and other difficulties. Lewis' main idea is due to Ramsey (1928, see Ramsey, 1978), and models a certain pragmatic feature: Think of theories as deductive systems in an axiomatic manner. One can have many true theories, some of them *more simple* than others, some of them *more informative* than others. Very simple theories have few and simple axioms, which they achieve at the cost of a poor informational content, while very rich theories are axiomatised with many complicated axioms. Now, according to Ramsey, laws are those sentences which are consequences in a deductive system which is axiomatised in the most simple way under the condition that the system captures everything worth knowing. In a sense, such a formulation does not only involve omniscience; as Lewis indicates, it appeals to God's standards for strength and informativeness and our standards for simplicity. Real scientific practice makes do with a trade-off between strength and simplicity. Accordingly, Lewis reformulates Ramsey's idea thus:

> a contingent generalization is a *law of nature* if and only if it appears as a theorem (or axiom) in each of the true deductive systems that achieves a best combination of simplicity and strength. (Lewis, 1973, p. 73)

This account explains many of the questions bedevilling the simple regularity theory. Why is one of two sentences with identical syntactic and semantic characteristics (equally general, in particular) a law of nature and the other not? They play different roles in systematising knowledge, differ in explanatory power or in bringing about simplicity of the whole construction. Why might the same sentence be a law in one theory but not in the other? Different theories about the same realm might be thought of as belonging to different possible worlds. Whether a sentence is a law of nature or not depends on which other sentences are true and take part in systematising the knowledge. Why are we inclined to consider generalised conditionals as candidates for laws of nature? All of them may play an important role in building up a deductive system.

The role of Lewis' conception of laws of nature in the discussion of whether differences in laws or differences in facts having greater bearing in determining overall differences in worlds is well known. Lewis himself argues that "small miracles" – that is, violations of laws of one world in another world – may be compatible with greater similarity of two worlds than a huge number of differences in facts do (Lewis, 1973, pp. 74ff.; Lewis, 1979). In order to understand why this question is important, one has to remember that Lewis' theory of counterfactuals is formulated semantically in terms of similarity between worlds. Lewis uses a comparative similarity relation which forces him to consider the question of what kind of differences make worlds more or less similar to others. Lewis also analyses causation in terms of counterfactuals, and in this way a theory of lawhood becomes crucial for counterfactual dependence and causality. Again, a well established and understandable notion of (comparative) similarity between worlds would allow for a better understanding of scientific practice. What does it mean to say that one theory is more true than another? If it is not true, it is false, so how are such claims to be understood? Lewis believes that an almost true theory consists of laws true of a world which is not so far (in the similarity metric) from ours (Lewis, 1986a, p. 24).

Blowing new life into the idea of an axiomatic theory of laws has found sympathy in certain circles, especially amongst those who wouldn't think of themselves primarily as philosophers of science. But two main objections can be raised against such an attempt. First, it would seem that we can only hope to be able to axiomatise some very general parts of physical theories. Most scientific knowledge cannot be handled within an axiomatic system. Second, it is only if one thinks in traditional analytical terms that one can hope to get an understanding of laws of nature based on an understanding of the structure of formal languages. Most philosophers of science would probably argue that laws exist objectively in nature, and this is not reflected in how we might characterise or organise our belief systems.

8. The necessity view

Some philosophers find the regularity view quite unsatisfactory, and this attitude often goes hand in hand with a strong metaphysical predilection. They believe that universal laws transcend experience; so an empirical analysis cannot provide a proper understanding of laws of nature. A metaphysical account is required. The positivists thought that a law of nature could be expressed in terms of a universal statement like 'All planets move in ellipses'. Although observation cannot prove such a claim since it is practically impossible to experience all planets in the universe, the expression says no more than what could in principle be observed if we had the power to scrutinise the entire cosmos. Thus, a statement like 'All planets move in ellipses' is appropriately confined to what goes on in the actual world. But, so the objection goes, there might be a regular correlation between two properties A (being a planet) and B (moving in an ellipse) by pure chance even if that the regularity holds without exception in the whole universe, whereas if it is a law that A is B, then it cannot be accidental that A is related to B. The claim that it is a law that all planets move in ellipses excludes the possibility that the uniformity of this regularity could be otherwise. The obvious answer is therefore that a law of nature does not connect two properties in a regular but possibly accidental way. There must be ways of interpreting the content of 'All planets move in ellipses' so that some kind of necessity is involved in relating being a planet and moving in an ellipse.

There are three different notions of necessity. The strongest is *logical necessity.* The usual definition claims that a sentence is logically necessary if the negation of the sentence entails a contradiction. This means that necessity holds in all possible worlds in terms of a definition. Laws of nature, however, do not seem to relate to anything by definition. Copernicus thought that the planets move in circles, and no one could correctly accuse him of contradicting himself. Moving in ellipses is not a part of our concept of a planet.

A weaker sense of necessity is that in which a law relates what co-exists in every possible world. We may call this *metaphysical necessity.* This sense is what Saul Kripke has in mind when he argues that terms like 'the morning star' and 'the evening star' rigidly refer to the same planet in every possible world. Think of planets and the eccentricity of their orbits. The metaphysical notion of necessity would imply that there is no possible world in which there are planets and they move in circles. But Copernicus might have been right had the world been different. The law of inertia might have been such that a body subject to no external force would not change its position (rather than its velocity). In this case a planet would move in circles. Thus metaphysical necessity also seems too strong.

The weakest sense of necessity is *natural* or *physical necessity.* The content of this notion is usually specified by saying that something is naturally necessary if, and only if, it holds in every possible world in which the laws of nature are valid. But obviously, such an explication is circular if an account of laws of nature is to be provided in terms of natural necessity. A philosopher who wants to appeal to natural necessity must avoid any charge of being incoherent.

The positivists believed that a law statement like 'All planets move in ellipses' could be expressed formally in an extensional language such as

(9) $\forall x(Ax \supset Bx)$.

But according to the necessity view, (9) does not suffice as a formal expression of the law in question. What we need is some sort of modal representation to express the lawful relation between being a planet and moving in ellipses. There are at least two ways of doing this: either a modal operator can be placed in front of the entire expression or in front of the consequent, thus:

(10) $\Box\forall x(Ax \supset Bx)$

(11) $\forall x(Ax \supset \Box Bx)$

Formula (10) states that the conditional is a necessary truth, i.e. that the consequent logically follows from the antecedent, whereas formula (11) means that for any object, if the antecedent is true of it, then the consequent is necessarily true of it. The readings are standard whenever the scope of the modal necessity operator differs. And the necessity operator is open to interpretation in each of the above senses of necessity. Which of these two statements, if any, does in fact reflect the nomic connection? Among the necessitarians there is little agreement about which of these interpretations gives us the correct understanding of laws of nature.

The regularity view on laws came under heavy fire in 1950s. William Kneale (1950) was one of the first to directly attack the positivist view of laws as universally quantified material implications. His criticism focuses on the unrealised physical possibilities. If laws of nature were merely to consist of uniform regularities, then laws could not explain the contrast between possible and impossible physical facts. But they can explain the difference. Although we have never seen a solid sphere of gold with a diameter of more than one mile, we believe that such a possibility cannot be logically ruled out on the basis of laws of nature. It is a physical possibility. We have never seen a sphere of enriched uranium 235 with a diameter of more than one mile, on the other hand, but in this case we are able to dismiss the idea as physically impossible because of the law of critical mass. In other words, we can formulate two universal statements concerning spheres of gold and spheres of uranium 235, to the effect that every such sphere is less than one mile in diameter. But only the one concerning the sphere of uranium 235 that is necessary.

In the attempt to find a replacement for the view that any universal statement represents a law of nature Kneale claimed that statements of laws of nature imply counterfactuals whereas universal material implications do not. In (Kneale, 1961) he argued that laws of nature are logically contingent but still necessary in the sense that they allow no alternative. A statement like 'All planets necessarily move in ellipses' excludes possible alternatives. So the kind of necessity he had in mind is not logical necessity, it seems, but "a generalisation which holds for all possible worlds of some kind" (p. 63). Some kind of nomological necessity is thus involved. Apparently, he regarded this modal component to be irreducible and not further explainable.

Karl Popper agreed with Kneale "that there exists a category of statements, the laws of nature, which are logically stronger than the corresponding universal statements" (Popper, 1959, p. 432). He accepted that laws of nature set certain limits to what is possible. If, say, it is a law that all planets move in ellipses, then it *would not be possible* for any planet to move in a circle. Moreover, a statement like 'All planets *necessarily* move in ellipses' belongs to the appropriate category of sentences that express this impossibility. Popper seemed to make no distinction between (10) and (11) as the correct formal representation of a law statement, however, since he says " 'If *a*, then necessarily *b*' holds if, and only if, 'If *a*, then *b*' is necessarily true" (Popper, 1959, p. 433).

What kind of necessity are we dealing with here, according to Popper? He himself maintained that a law of nature contains a natural or physical necessity that prohibits planets from moving in circles. It is conceivable, he said, that they do move in circles in some possible worlds but not in those where the law holds. The delicate question therefore is how he defines natural necessity. In *The Logic of Scientific Discovery* he suggested the following definition:

> (N^o) A statement may be said to be naturally or physically necessary if, and only if, it is deducible from a statement function which is satisfied in all worlds that differ from our world, if at all, only with respect to initial conditions. (Popper, 1959, p. 433)

Unfortunately, as we shall see, this definition did not enable Popper to avoid the threat of circularity. He even realised this himself:

> Nevertheless, the phrase in (N^o) 'all worlds which differ (if at all) from our world only with respect to the initial conditions' undoubtedly contains implicitly the idea of laws of nature. What we mean is 'all worlds which have the same structure – or the same natural laws – as our own world'. In so far as our *definiens* contains implicitly the idea of laws of nature, (N^o) may be said to be circular. But all definitions must be circular *in this sense* . . . (Popper, 1959, p. 435)

First, we should notice that the first quotation describes a statement as being naturally or physically necessary. Popper wrote as if *laws of nature* are statements. He also said that he regarded 'necessary' as a mere word, a label for distinguishing the universality of laws from 'accidental universality'. Any other label could be used since the idea is not much connected with logical necessity. And Popper confirmed his agreement with "the spirit of Wittgenstein's paraphrase of Hume: 'A necessity for one thing to happen because another has happened does not exist. There is only logical necessity'" (Popper, 1959, p. 438). There are indeed good reasons to be puzzled by Popper's confused ways of expressing himself. Kneale (1961) correctly accused him of being inconsistent. The kind of necessity Popper had in mind cannot, if it is to make any sense, be a mere *de dicto* form of necessity. It does not make sense to claim that *physical* necessity holds among sentences but not in the real world. Such a notion requires that the laws of nature are also valid with respect to states of affairs other than the actual. The correct way of talking for someone who holds that laws of physics are physically necessary would be to say, "A statement may be said to express a natural or physical necessity ...". As an empiricist he cannot both be terrified by essentialism and, at the same time, operate with a notion of natural or physical necessity defined in terms of possible worlds.

Second, the first quotation characterises different possible worlds in terms of initial conditions and not directly in terms of physical laws. The consequence, however, is the same. Any talk of initial conditions makes sense only with respect to a given set of law statements. Natural or physical necessity is being defined in terms of laws of nature and, therefore, laws of nature cannot be defined with the help of natural or physical necessity. Popper does not think so because he holds that all definitions are circular in this sense. It may be true in cases where we consider a huge segment of a certain vocabulary and do not allow primitive terms in this vocabulary. But the circle in question is based on a very small segment, and Popper did not, it seems, hold the notion of law to be primitive. The upshot is that it is rather vacuous to distinguish universal statements from law statements in virtue of an appeal to natural or physical necessity since the meaning of the latter cannot be understood independently of the meaning of the former.

In the late 1970s David Armstrong (1978), Fred Dretske (1977), and Michael Tooley (1977) suggested that laws of nature are relations holding between universals. To say that it is a law of nature that Fs are Gs means that F necessitates G. The basic idea behind this view is that a law itself is not a necessity but accounts for necessity. Armstrong's book *What is a Law of Nature?* (Armstrong, 1983) is the most well-developed analysis of the three, and his account will be taken up here.

Like Kneale and Popper before him, Armstrong rejects the idea that a statement such as 'It is a law that all planets move in ellipses' is equivalent to 'All planets moves in ellipses'. Instead he considers it to be equivalent to 'It is physically necessary that all planets move in ellipses' (p. 77). This immediately raises two questions. How should this explication be understood formally, and what kind of necessity is involved?

If we take the first question first, it seems clear that Armstrong takes the necessity term to have the broadest possible scope. We should therefore expect him to take (10) to provide us with the correct formal representation of 'It is a law that Fs are Gs'. But he argues that even if (10) corresponds logically to what it means to be a law of nature, it doesn't do so metaphysically. Placing the necessity operator in front of (9) is, he says, "merely a technical solution unaccompanied by metaphysical insight" (Armstrong, 1983, p. 87). Moreover, he believes that (10) also faces internal problems in view of the connective – the material implication raises the well-known Paradoxes of Confirmation (pp. 87–88). As an alternative formulation, he proposes

(12) $N(F, G)$,

where (12) implies (9), *mutatis mutandis*, but not vice versa. So 'N' replaces '$\supset$' as the logical connective. Armstrong considers N to be a relation of nomic necessitation that connects two universals F and G. Moreover, $N(F, G)$ is itself a universal, a second-order universal, and Armstrong says that seeing this helps us to accept that N is a primitive relation. 'N' stands for "a real, irreducible, relation, a particular species of the necessitation relation, holding between the universals F and G" (Armstrong, 1983, p. 97).

So how should (12) be interpreted? Armstrong is clear on this point. It should be read as 'Something's being F necessitates that same something's being G, in virtue of the universals F and G' rather than 'For all x, x being F necessitates that x is G'. (Armstrong, 1983, p. 96) The former statement reflects the idea that the same relation of necessitation holds between sorts of states of affairs. The latter statement expresses, according to Armstrong, merely a more advanced form of the regularity view because it is assumed that a singular necessitation holds between particular states of affairs. Hence, it is equivalent to the formula:

(13) $\forall x N(Fx, Gx)$.

Their dependency is such that (12) entails (13), which again entails (9), but the reverse entailments do not hold.

This brings us to the second question. How can we identify the relation of necessitation? Armstrong believes that there are purely singular relations of necessitation without laws, and some of them are cases of singular causation which exist independently of a law. But can we be sure that singular necessitation corresponds to something in the real world since it does not correspond to the fact that a being F co-occurs with a being G? How can we identify the state of affairs that corresponds to the necessitation relation other than by invoking a claim such as effects follow their causes? Is it possible to identify the truth-maker of a singular necessitation statement independently of the claim itself? Even if we get a satisfactory answer to these questions, we cannot, without further arguments, use it to explain why one type of state of affairs (the universal F) necessitates another type of state of affairs (universal G) because (12) is not merely a generalisation of singular necessitation. The fact is that Armstrong does not tell us how to identify necessitation.

Are laws of nature necessary or contingent? Armstrong excludes strong necessity, which he represents as

(14) $\Box N(F, G)$.

This formula is analogous to (11). He takes it to mean that in every possible world the relation N relates F and G. So his understanding of the necessity operator is that of logical necessity. But, as we have seen, the necessity operator has alternative possible interpretations. Nevertheless, he rejects the proposal because this would make not only laws but also universals *necessary beings* and there are, he argues, definitely worlds without Fs and Gs (Armstrong, 1983, p. 164). Hence F and G are contingent beings. Moreover, laws of nature might have been different from what they are, and they cannot therefore be strongly necessary.

Armstrong also discusses another formulation as a candidate for 'It is a law that Fs are Gs' which he takes to express a weaker form of necessity. Here the formulation combines the contingency of universals with the necessity of laws:

(15) $\Box($ the universal F exists $\supset N(F, G))$.

Apart from the necessity operator in the front, (15) bears a certain resemblance to (11). Spelled out in ordinary terms, it says that in all those worlds which contain the universal F, it is a law that Fs are Gs. This view requires the existence of irreducible powers, Armstrong argues, but it does not fit well with his hopes of an actualist metaphysics.

> Although I do not believe in the literal reality of possible worlds, or even in the literal reality of ways things might have been but are not, I know of no way to argue the question before us except by considering possible worlds. It may be that the necessary/contingent distinction is tied to a metaphysics which recognizes possibility as a real something wider than actuality. If this could be shown, then my inclination would be to abandon the necessary/contingent distinction and declare our present question about the status of the laws of nature unreal. But I cling to the hope that an account of 'possible worlds' can be given which does not assume the existence of *possibilia*. (Armstrong, 1983, p. 163)

Whether or not such an actualist account can be given remains to been seen. For laws are logically contingent, according to Armstrong, but also physically necessary (Armstrong, 1983, p. 77). He also mentions it as a contingent necessity that being F necessitates being G. One philosopher, as we shall see later, has bitten the bullet and declared not only that possible worlds are fictions, but also that laws of nature are unreal.

9. Conventionalism

A very different approach to understanding laws of nature is found among conventionalists. They focus on the idea that fundamental law statements, like Newton's three laws of motion, resist falsification. Henri

Poincaré, the father of conventionalism, emphasised this in particular by saying that an empirical law is always subject to revision but no one seriously believes that any of Newton's laws will be abandoned or amended. Why is this so?

Explaining it, we should notice that Poincaré distinguished, in his Preface to *Science and Hypothesis*, between three kinds of laws, or rather three kinds of hypotheses expressing such laws. These are (1) *experimental laws*, i.e. hypotheses that "are verifiable, and when once confirmed by experiment become truths of great fertility"; (2) *principles*, i.e. hypotheses "useful to us in fixing our ideas"; and (3) *definitions*, i.e. hypotheses "that are hypotheses only in appearance, and reduce to definitions or to conventions in disguise". Although he did not at this place use the terms "experimental laws", "principles" and "definitions" himself, they can be found elsewhere in the text. Furthermore, he believed that the fundamental laws of classical mechanics and the principle of the conservation of energy belong to the category of principles and definitions.

Principles are conventions or definitions in disguise. They are deduced (generalised) from experimental laws in the sense that these laws have been elevated to principles, which the scientific mind attributes an absolute status for the time being. Such conventions are not absolutely arbitrary. We accept them because certain experiments have shown us that they are convenient. Also Poincaré gave an explanation of how a law can become a principle. First, we have a hypothesis that expresses a relation between "two real terms" A and B. It does not state a rigorous truth but only an approximation. Second, we arbitrarily introduce an intermediate term, C, which he characterised as more or less imaginary. The term C is therefore a result of abstraction and idealisation. Now C is "*by definition* that which has with A *exactly* the relation expressed by the law". And he continued:

> So our law is decomposed into an absolute and rigorous principle which expresses the relation of A to C, and an approximate experimental and revisable law which expresses the relation of C to B. But it is clear that however far this decomposition may be carried, laws will always remain. (Poincaré, 1952, p. 139)

An example is the experimental law that the sum of the kinetic and potential energy, $T + U$, is constant. Kinetic energy is proportional to the square of the velocity and the potential energy is independent of the velocities. We might think that the conservation of energy could be gained experimentally by adding the internal molecular energy to the sum of kinetic and potential energy $T + U + Q$ where Q is independent of the position and velocity. But Poincaré argued that this is not possible since the internal molecular energy is not only dependent on their internal state. The electrostatic energy of electric charges depends on both their positions and their velocities. Therefore, the three terms in $T + U + Q$ are not absolutely distinct. So it is not only true that $T + U + Q$ is constant; the same is true of any function $\varphi(T + U + Q)$ whatsoever, and "among those functions that remain constant there is not one which can rigorously be placed in" the particular formula of three distinct terms. The consequence is that the conservation of energy can never be equivalent to the empirical law that $T + U + Q$ is constant.

Poincaré did not claim that, say, Newton's laws of motion couldn't be disconfirmed because these statements are a priori true. If, say, the law of inertia were imposed on us a priori, it would be impossible to understand why the Greeks never got it right. Furthermore, one cannot argue that velocity does not change unless acted upon by saying that this is the only law compatible with the principle of sufficient reason. For the world might well have been different in such a way that it is a law of nature that the position or the acceleration of a body would be unchanged if it were not acted upon by a force.

But neither does the law of inertia express an empirical fact because it is impossible to conceive evidence against it. It is impossible, Poincaré says, to make experiments on bodies on which no forces act, and even if it were possible, we would have no means to know that no forces were acting on such a body. Thus, he called the law of inertia *the principle of inertia*, indicating that such a statement should be considered basic or superior in our thinking, expressing a rule which our thoughts obey in the description of nature. We do not have to adhere to them, but we do for reasons of convenience since it is "useful to us in fixing our ideas".

Poincaré called Newton's second law of motion the law of acceleration. We can measure acceleration, he argued, but we cannot measure force or mass, and we don't know what they are. We have one equation with two unknowns. If we claim, for instance, that "force is the cause of motion, we are talking metaphysics" (Poincaré, 1952, p. 98) because we cannot establish what makes a force F equal to a force F^*. This is necessary if we are to check that the same numerical forces produce the same numerical accelerations when applied to the same numerical masses. We may believe that it is possible to measure masses independently by their weight, but the weight of the same mass varies with respect to gravitation. We need a third definition, Newton's third law, which equates action and reaction. But, as Poincaré pointed out, "we are compelled to bring into our definition of the equality of two forces the principle of equality of action and reaction; *hence this principle can no longer be regarded as an experimental law but only as a definition*" (Poincaré, 1952, p. 100).

The third law then defines what it means for a force to be equal to another. Assume two bodies, A and B, act on each other. The acceleration of A times its mass is then said to be equal to the action of B on A, and in the same way the acceleration of B times its mass is equal to the action of A on B. But two bodies are never alone and cannot be abstracted from the rest of the world. We have no means of distinguishing the action between A and B from the acceleration due to all other bodies, and this makes the decomposition of the various central forces involved impossible. The upshot is that "masses are coefficients which it is found convenient to introduce into calculations" (Poincaré, 1952, p. 103).

So Poincaré believed that the law of acceleration and the decomposition of forces were conventions, although not arbitrary conventions, because their adoption was based on experiments. Some philosophers have similarly argued that the basic laws of nature should be regarded as definitions, whereas others have opted for a view according to which it is the specific situation that determines whether laws of nature should be considered definitions or empirical hypotheses.

10. The contextual view

One of Poincaré's contemporary countrymen, Pierre Duhem, briefly discussed conventionalism in his *The Aim and Structure of the Physical Theory* which was published a year after *Science and Hypothesis.* According to one widespread interpretation, he himself regarded laws of physics as neither true nor false but approximate, in virtue of which he says they are relative, and they are provisional because they are symbolic representations. On this interpretation, he was sympathetic to the insight behind the conventionalist view that the fundamental laws such as Newton's laws of motion may act like definitions and therefore be almost immune to falsification. But he also stated that such confidence in a law of nature is not "analogous to the certainty that a mathematical definition draws from its very essence" (Duhem, 1906, p. 211). If the result of an experiment disagrees with a certain theory, we do not know which part of the symbolic representation that has to be rejected but we know that some part must be discharged. A single experiment can never condemn a hypothesis in isolation, it can only question a whole theoretical system of laws. He also thought that, whenever physicists of a certain epoch look for possible modifications, there is always a certain number of laws which they agree to accept without further test because they consider them beyond dispute. But Duhem warned against believing that physicists are forced to act in this way because of logical necessity. They do so because to act otherwise would be irrational.

So Duhem accepted that laws might be treated as conventions.[6] He wanted, however, to warn against considerations which would treat them as analytic truths. He concluded:

> ...we must really guard ourselves against believing forever warranted those hypotheses which have become universally adopted conventions, and whose certainty seems to break through experimental contradiction by throwing the latter back on more doubtful assumptions. The history of physics shows us that very often the human mind has been led to overthrow such principles completely, though they have been regarded by common consent for centuries as inviolable axioms, and to rebuild its physical theories on new hypotheses. (Duhem, 1906, p. 212)

Duhem knew the history of physics better than anyone did. He therefore spoke with the insight of an active scientist as well as that of a historian of science. But one should also remember that Poincaré knew better than most other physicists that conventions could be discarded as inadequate since he, and Einstein, took the lead in the revolution of relativity overthrowing the Newtonian physics. So Poincaré was very open-minded concerning the exchange of one set of conventions with an alternative formulation of them.

When discussing fundamental laws as stipulative definitions a distinction should be kept in mind. The notion of definition is very sensitive to a context and relative to a set of interest. Hence, scientific laws can be divided into definitions, principles and experimental laws without any of these being completely sacrosanct whenever a theory runs into trouble. A law being a definition means being analytic-in-a-theory and does not mean being analytically true. The difference is that an analytic statement is true by virtue of synonymous meaning, and is traditionally regarded as an example of priori knowledge, whereas definitions as analytic-in-a-theory are meaning-constitutive for that theory but their truth-values are empirically determined. In fact it is not definitions themselves that are true or false, but rather the truth-values of concrete sentences expressed according to these definitions. Definitions as meaning-constitutive stipulations are more or less adequate. Laws that are regarded as stipulative definitions will therefore resist empirical revision much longer than experimental laws.

[6] One of us has questioned the antirealist interpretation of Duhem and argued against the conventionalist interpretation (Needham 1998, 2000). On this view, Duhem's argument against Poincaré was to the effect that the holistic nature of testing doesn't allow for any distinction in principle between ordinary laws and those like the principles of mechanics which Poincaré held to be conventions.

Ideas similar to those of Duhem prompted philosophers, like Russell Norwood Hanson (1958) and Thomas B. Kuhn (1962/70), to propose a highly context-dependent notion of fundamental laws. Russell Hanson argued that expressions like 'Newton's second law' and 'The law of gravitation' should be considered as 'umbrella-titles' since the formula '$F = m\mathrm{d}^2x/\mathrm{d}t^2$' and '$F = \mathrm{k}(m_1m_2)/r^2$' can be used as definitions, a priori expressions, heuristic principles, empirical hypothesis, rules of inferences, etc. (Hanson, 1958, p. 112). Which of these distinct uses a physicist would take to be relevant depends on the concrete situation and the general purpose he wants to pursue. In the discovery of the non-visible planets, Uranus, Neptune and Pluto, Newton's law of gravitation was treated more as a definition than anything else. Russell Hanson mentions an example where Newton's laws were put to test as an empirical hypothesis, however. In 1784 the English physicist George Atwood wanted to show that Bernoullis' and Leibniz's attacks on the Newtonian mechanics were ill founded, and he conducted an experiment to prove that Newton's laws were consistent with the facts.

Kuhn held a view similar to Russell Hanson's. He distinguished between exemplars and symbolic generalisations. Exemplars are the application of the symbolic generalisations to specific types of situation in accordance with so-called standard models. However, much like Duhem, he thought of Newton's laws, Coulomb's Law, Ohm's law, and Maxwell's equations as symbolic generalisations rather than empirical hypotheses because in their most general formulation they don't say anything about concrete physical systems. They are kinds of abstract schemata which physicists have to concretise before they can apply them to particular cases. According to Kuhn, "they function in part as laws but also in part as definitions of some of the symbols they deploy. Furthermore, the balance between their inseparable legislative and definitional force shifts over time" (Kuhn, 1962, p. 183). In other words, the historical context has an important say in the way physicists would treat symbolic generalisations. In the beginning physicists would look at them more as empirical hypotheses; later on, when these have become parts of a successful paradigm, more as definitions. He summarised his opinion in the following way: "I currently suspect that all revolutions involve, among other things, the abandonment of generalizations that force of which had previously been in some part that of tautologies" (Kuhn, 1962, pp. 183-184).

The contextual view sees certain basic laws as symbolic representations whose function is determined by the role which physicists assign to them in a certain historical situation. As law-schemas they are idealisations that do not apply directly to a physical system, but must be mediated through models of application. Therefore this view entertains a certain kinship to the *ceteris paribus* account of laws. Many will, nonetheless, feel a strong reservation against contextualism because of, as they see it, its ontological shortcomings. For contextualists must hold that law statements do not carry a literal meaning since the linguistic function of law statements depends on the context in which they are asserted and their ontological status is relative to this function. Hence there are no real matters of fact corresponding to a law statement and described in terms that are literally true or false. Theories cannot be true or false, as advocated by Duhem, and some kinds of instrumentalism or constructive antirealism must follow.

11. What if there are no laws?

The contextual view threatens to render a substantial notion of laws of nature obsolete and allow laws once again to be reduced to regularities. There are no laws, on this view, only restricted regularities. This is Bas van Fraassen's position. In his book *Laws and Symmetry* he takes issue with the traditional, and metaphysically based, notion of laws as it traces back to Descartes in the seventeenth century. Such a notion no longer belongs to natural science, he maintains. Today it figures only in philosophical writings, where the aim of science is held to be the discovery of laws of nature, which are therefore a central ontological concern for the philosophy of science.

Within the metaphysical tradition, van Fraassen recognises two arguments for a robust notion of laws. One argues that there are pervasive and stable regularities in nature, but that no such uniformity can exist by chance. Why should there be such regularities if is wasn't for the existence of laws of nature? The other argument holds that denying that there are reasons for a regularity leads to scepticism because without laws of nature it is not rational to expect the future to be like the past. But he concludes:

> I can quite consistently say that all bodies maintain their velocities unless acted upon, and add that this is just the way things are. That is consistent; it asserts a regularity and denies that there is some deeper reason to be found. It would be strange and misleading to express this opinion by saying that this is the way things are by chance. But that just shows that the phrase 'by chance' is tortured if we equate it to 'for no reason'. (van Fraassen, 1989, p. 21)

Undoubtedly, the metaphysician will object that laws and truths in general differ, that there are criteria, which a law of nature can meet, but which a mere regularity cannot, such as universality, necessity, and objectivity. van Fraassen discusses these criteria one by one. He rebuts the demand for universality by quoting an example from Reichenbach and Hempel:

1 All solid spheres of enriched uranium (U235) have a diameter of less than one mile

2 All solid spheres of gold (Au) have a diameter of less than one mile.

Although both statements are true, van Fraassen agrees that it is only the first expresses a putative law because of uranium's critical mass; the second states an accidental fact. Nevertheless, both are universal to the same degree. Moreover, they are in general syntactically and semantically indistinguishable, and it is therefore impossible to identify laws as the true "law-like" statements. We cannot separate a law statement from an accidental statement in virtue of language. He also points out that it is almost impossible to specify generality of content, and that it might even be the case that universality is not a requirement for being a law.

Necessity is also a general feature associated with laws of nature. We usually think that what participates in such a law must obey it. van Fraassen distinguishes between four kinds of necessity: inference, intensionality, *the necessity bestowed* and *the necessity inherited.* Inference and intensionality are matters of logic, which is a matter of what conclusions follow from premises. Those are the only kinds of necessity empiricists can accept. The notion of necessity bestowed, according to which "It is a law of nature that A is true" is considered equivalent to "It is necessary that A is true", is a notion that they reject. Even more so, they reject a stronger notion according to which the necessity is inherent in the laws themselves. Such features of necessity are not immediately perceptible, and no inference such as inference to the best explanation has the strength to take us from the sort of facts that actually can be observed to a claim that unobservable facts of necessity provide us with the best explanation of regularities.

Lawfulness is commonly linked to counterfactuals in so far as giving warrant for counterfactual conditionals is regarded as a criterion of law. In the mid-1940s Nelson Goodman and Roderick Chisholm realised that counterfactuals do not reflect the same principles of reasoning that hold for strict or necessary conditionals. But van Fraassen thinks that the semantic analysis of counterfactuals, which Robert Stalnaker and David Lewis carried out in the late-1960s, shows that the behaviour of counterfactual conditionals deviates from that of strict ones because of context-dependence. This means that the most interesting counterfactuals do not derive from necessities alone "but also from some contextually fixed factual considerations". van Fraassen denies, however, that science itself implies any interesting counterfactuals. So he argues that if laws did in fact imply such conditionals then they would have to be indexical statements, which conflicts with the idea that they have a purely objective content.

Apart from having to address these criteria of lawfulness, van Fraassen believes that any philosophical account of laws of nature must face two major problems, which he calls *the problem of inference* and *the problem of identification*. An easy solution to one of them creates unavoidable difficulties for the other. Take the question of inference. It is a reasonable requirement of any account to say that if it is a law that A, then it implies that A. A simple solution that makes such an inference possible is to equate *It is a law that A* with *It is necessary that A*, relying on the logical principle that necessity implies actuality. But, as van Fraassen points out, we now have a problem of identifying what sort of facts makes such a claim true. If the problem is met by maintaining that necessity is itself a primitive fact, however, then it is not clear why necessity should be thought to include actuality. Conversely, if we first identify those actual regularities that seem to lead us to a claim that it is a law that A, there is little or no room for an inference that there is a law that A, and therefore that A because we are haunted by the problem of induction.

The only tenable solution, according to van Fraassen, is to claim that there are regularities in the world but no laws. His constructive empiricism is well known. The central tenet of this view, as presented in his *The Scientific Image* (1980), is that the aim of scientific theories is not to yield true descriptions of the world, but to give us empirically adequate descriptions of phenomena. He also denies that we accept scientific theories because we believe that they are true. Rather, we accept them because we believe that they are empirically adequate. A description is said to be empirically adequate if it is true with respect to what can be observed, and only to what can be observed. As a consequence, van Fraassen views with great suspicion any position that tries to vindicate the existence of hypothetical entities or lawful relations, which cannot be seen by the naked eye. Whether the argument focuses on the reality of unobservable entities or the reality of laws, it makes, unjustifiably, a leap from what can be actually known to be to what may possibly be. Empiricists, like van Fraassen, have always had their doubts about *real* possibilities other than actualities.

Most other accounts of natural laws introduce modality as something in virtue of which laws can be characterised in order to distinguish them from mere regularities. A constructive empiricist, however, cannot accept such accounts in terms of *de re* modalities since necessities and possibilities are inaccessible to the faculty of perception. As van Fraassen puts it in *Laws and Symmetry*: "From an empiricist point of view, there are besides relations among actual matters of fact, only relations among words and ideas. Yet causal and modal locutions appear to introduce relations among possibilities, relations of the actual to the possible" (van Fraassen, 1980, p. 213). Rather than being part of nature, modalities are part of language, and a philosophical explication of modality is to be part of a theory of meaning. Thus, van Fraassen solves the dilemma by arguing that there are no laws.

12. The *ceteris paribus* View

A large part of the recent literature on laws of nature, and many of the essays in the present volume, relate directly or indirectly to Nancy Cartwright's works. Discussion centres on her characterisations of fundamental laws either as false or as being *ceteris paribus* laws. In her later works she combines these features with an understanding akin to the necessity view. As she says, "a law of nature is a necessary regular association between properties antecedently regarded as OK" (Cartwright, 1999, p. 49). However, as discussions in this book show, nothing forces a proponent of the *ceteris paribus* view to accept Cartwright's view on capacities. The *ceteris paribus* view can be combined with conventionalism, contextualism and Duhemianism.

A *ceteris paribus* clause expresses that there are some circumstances which must be fulfilled in order for the statement it makes conditional to be true. By adding the clause, the universal applicability of the statement in question is narrowed. Nancy Cartwright does not distinguish between different types of *ceteris paribus* clause. A distinction can be made, however, between those clauses that require the fulfilling of certain abstract or ideal circumstances and those that require the realisation of certain factual or concrete circumstances. Newton's law of gravitation illustrates the first type. It is true only if the bodies to which it is applied are regarded as point masses and only if there are two of them. But such conditions are not of this world. It is therefore often said that Newton's law of gravitation is only true of a model. The other type is exemplified by causal statements. For instance, a causal claim like striking a match makes it light is true just in case certain concrete circumstances are fulfilled: the presence of oxygen, the match is dry, and the right amount of friction is applied when pressing the match against the sulphur, etc. Consequently, causal statements are true of the world but fundamental laws are not.

Cartwright's influence on the discussion is partially explained by her holding a strong and somewhat counterintuitive sounding thesis:

> ...the fundamental laws of physics do not describe true facts about reality. Rendered as descriptions of facts, they are false; amended to be true, they lose their fundamental, explanatory force. (Cartwright, 1983, p. 54)

How should one understand such a challenge to common-sense ideology? Usually, fundamental laws are taken to explain, and to support epistemologically, phenomenological laws (of physics and other sciences). On this view, doubting fundamental laws might be motivated from the traditional antirealist position that focuses on the theoretical terms use in the formulations of fundamental laws. These have to be connected to entities figuring in phenomenological laws. Worries about the status of theoretical terms, about their relationship to experience or holding outright that no claims whatsoever relate to reality, lie behind traditional antirealist positions. Cartwright, on the other hand, does not worry about theoretical terms and nor does she believe that statements can represent facts of nature. Although she first dubbed her position antirealist, later on she singles out fundamentalism as the real enemy (Cartwright, 1999, Ch. 1). The failure lies in the nature of explanation.

The division of laws of nature just mentioned into phenomenological laws and fundamental laws becomes clear from Cartwright's examples. Most biological laws or engineering rules are phenomenological, whereas Coulomb's law and the law of gravitation are fundamental. The point of her arguments consists in an observation about the role fundamental laws must play in scientific practice. In order to be true, they have to describe the behaviour of entities, but in order to explain, the "composite" forces have to be disassembled into more elementary parts, for it is these that are described by the fundamental laws. Cartwright denies that there is something like a law of composition of forces comparable to the rules of vector addition which can be applied to take them apart and put them together. A charged massive body moves neither according to the law of gravitation, nor according to Coulomb's law, nor according to the laws of a third force which would be a resultant of the two aforementioned forces. Fundamental laws are designed to explain phenomenological laws; they are tools to systematise them and it one of their essential features that they explain more than one of them. But precisely this feature prevents them from being true. If they were true, they could not explain; if they explain, they cannot be literally true.[7]

This more critical part of Cartwright's conception amounts to the idea that phenomenological laws – which reflect things as they actually are in nature – cannot be derived from fundamental laws by laws of logic. Neither are fundamental laws true abstractions of phenomenological laws, speaking about the same things in a more general and abstract manner. With the help of many examples, Cartwright argues that the real relationship between the two types of laws has to be understood to be created by a long and complicated process of approximations and emendations. If she is right, the problem consists in a misunderstanding about how the laws of physics explain. Her first suggestion for solving this problem she called a "simulacrum account". This account appeals to two senses in which the term "realistic" might be applied. In a first sense, physical theories may be realistically interpreted if they give a close enough description of what happens in reality. In this sense, the more realistic the theory, the fewer *ceteris paribus* conditions or fewer physically unrealised assumptions are involved. In a second sense, physical theories may be realistically interpreted if they explain what happens in the mathematical apparatus. Physical models may be realistic in many respects. One might wish to calculate a functional relationship with great precision, one might wish to understand a causal relationship in detail, or one might be interested in completely different properties. A physical model is a simulacrum insofar as it shares certain properties with the modelled part of the world (to a greater or lesser extent), although it still has many other properties which are not essential and perhaps even completely arbitrary: "A model is a work of fiction" (Cartwright, 1983, p. 153). Usually, Cartwright criticises, many physicists conclude that even the "properties of convenience" of a model which fits a simple case must be there when it is applied to much more complicated cases. Unfortunately, she is not able develop the simulacrum account in a formal manner, but she describes it as follows:

> It [*the simulacrum account*] says that we lay out a model, and within the model we "derive" various laws which match more or less well with bits of phenomenological behaviour. But even inside the model, derivation is not what the D-N account would have it be, and I do not have any clear alternative. (Cartwright, 1983, p. 161)

[7]That does not mean that there aren't any true explanations. Accepting, for instance, the causal explanation of a fact of nature means simply accepting the cause – even if this involves theoretical terms.

Nature is complex through and through, so simplicity is gained only at the cost of misrepresentation, particularly of misrepresentation of singular causal processes. Since grasping causation seems to be crucial for understanding what is going on "inside the model", Cartwright turns to the problem by finding an account of causality.

Why do causes increase the probability of their effects? Because the single cause, Cartwright says, has the capacity to do so. She even thinks that the metaphysics of capacities provides an answer to questions such as why theoretical laws are applicable in different situations. If causes have the capacity to bring about a certain effect, then they can be expected to carry that capacity from one situation to another. Even the fact that causes increase the probability of their effects *in all causally homogeneous circumstances* shows that capacities exist:

> Just as laws constrain relations between matters of fact, capacities constrain relations among laws. A property carries its capacities with it, from situation to situation. That means that, where capacities are at work, ... one can infer from one causal law directly to another, without ever having to do more tests. (Cartwright, 1989, p. 146)

Cartwright speaks about "modal levels" where the claims at the higher level constrain what structure the set of claims at the lower level can have (Cartwright, 1989, p. 160):

Levels of Modality:	Ascriptions of capacity Causal Laws Functional and probabilistic laws
Non-modal Level:	Occurrent regularities

In that picture, capacities play a special role. They are not only modalities; they are in the world. Cartwright would have us believe that nature selects several capacities that different factors carry, and it sets the rules for their interplay. From this, causal laws as well as everything under them (see table) gets shaped. Speaking metaphysically, capacities are natures, or essences; they exist, whereas associations are epiphenomena, and ontologically secondary.

Still, the problem which appeared right at the beginning of Cartwright's discussion remains. Basic facts about phenomenological regularities are reconstructed well enough in low-level laws (as in biology, technology or econometrics). Higher-level laws are designed to explain these phenomenological laws rather than the behaviour of things in the world. Cartwright was able to show that this is done, not by logical deduction and not by deductive nomological reasoning, but rather by putting certain constraints on lower levels. It remains to be explained how this is realised in the case of capacities with respect to fundamental laws. If Cartwright could answer this question she would have solved the problem mentioned at the beginning of this section: How do laws explain?

The key notion of Cartwright's conception of where laws of nature come from is that of a "nomological machine":

> It is a fixed (enough) arrangement of components, or factors, with stable (enough) capacities that in the right sort of stable (enough) environment will, with repeated operation, give rise to the kind of regular behaviour that we represent in our scientific laws. (Cartwright, 1999, p. 50)

A nomological machine is an arrangement of certain capacities, which can be rearranged and collected together in other such machines. Thus, Kepler's law about the motion of the planet Mars along an elliptic orbit is really a law about physical bodies moving in time and space. Newton's law of gravitation does not speak about bodies moving in time and space; rather, it introduces a capacity: in the right circumstances, a force has the capacity to change the motion of a body. This capacity may, of course, be used in other nomological machines, too. In any case, it will shape the character of the regularities we expect to hold.

As can be seen from the example, nomological machines are like models in some respects. They need the right parts – capacities – which have to be arranged in the appropriate manner. They have to be shielded; that is, they will give the results desired only in the stable circumstances mentioned. Hence, all laws stemming from a nomological machine will come with *ceteris paribus* conditions. The very idea of compositionality, which is part of what is involved in the idea of explanation (the movement's characteristics x, y are explained by one reason, y, z by others) is build into the concept of a nomological machine, and since capacities are tendencies or propensities (as distinct from Carnapian dispositions), they can explain without real deductive force.

13. Summary of the papers

The present volume collects papers on the essential character of laws of nature. A variety of conceptions are presented by a dozen authors. Together they give a wide-ranging overview of the state-of-the-art of contemporary discussion on the subject.

Mauro Dorato's paper "Why are (most) laws of nature mathematical?" concentrates on measurement too. He addresses Eugene Wigner's question about the unreasonable effectiveness of mathematics in the natural sciences. Dismissing the so-called "software approach to laws", Dorato arrives at the conclusion that laws of nature are mathematical because they are expressed in a mathematical form isomorphic to the relational structure of the respective natural systems. The essence of this isomorphism, however, emerges from the fact that many mathematical concepts do have empirical origins.

There is a prominent view of laws of nature as linguistic conventions or rules of language, advocated, for example, by Henri Poincaré. Defending a variant of this approach in "How nature makes sense", Jan Faye distinguishes between two sorts of laws: theoretical and causal. Methodologically, theoretical laws precede causal laws – we need the first to formulate the second. He argues, against Nancy Cartwright and others, that theoretical laws don't involve *ceteris paribus* clauses, but causal laws do. It is causal laws that allow "nature's sense" to emerge. According to Faye, causes are grounded in the way we see nature, so causal necessity is bound to possible experience. Theoretical laws, on the other hand, have no descriptive content and are therefore conventional in a sense.

Igor Hanzel analyses in great detail the evolution of the view on laws of one of the most influential contemporary philosophers of science against the approach presented by one of today's most interesting philosophers of science. In his "Nancy Cartwright on laws as lies and as capacity claims" Hanzel demonstrates how Cartwright's lack of knowledge of Leszek Nowak's work predetermined the style of her own considerations. In effect, Cartwright's later project, seeing laws as capacities, also seems problematic in the face of, for example, theoretical physics.

Henrik Hållsten sets out to defend explanatory deductivism. To do so he has to come to terms with probabilistic causes: granting them explanatory force leads to unwanted consequences. In his "The explanatory virtues of probabilistic causal laws" Hållsten argues that a probabilistic cause (if there is such a thing at all) would be neither necessary nor sufficient for its effect. What we have in such a case, essentially, are chances. These chances do not add up to an intuitive cause, whereas the degrees of our rational beliefs do add up.

In his "The nature of natural laws" Lars Göran Johansson discusses the definition of the concept of a natural law. He argues that there are various forms of such laws, each requiring different treatment. Some laws express quantitative relations. These are derivable from a set of fundamental laws, i.e. of implicit definitions of the predicates involved. Some other laws are conservation principles. Johansson identifies them as consequences of objectivity demands on the descriptions of physical systems. Some laws are neither of the above and Johansson's analysis does not touch on them. It does, however, cover quite a number of important cases of natural law.

Geert Keil argues that there are no strict laws of succession, i.e. universally quantified conditionals saying that any event of type *c* is followed by an event of type *e*. Because all such statements are subject to exceptions, they are compromised as laws. However, the title, "How the *ceteris paribus* laws of physics lie", clearly expresses the authors opinion that this doesn't apply to all physical laws. In the course of his discussion, Keil presents a very interesting overview of various ways to understand the term "*ceteris paribus*".

Challenging the empiricist tradition, Max Kistler defends the necessity of (at least some) laws of nature. He does so by using subtle metaphysical arguments. His "Necessary Laws" ends with a clear-cut conclusion: laws are second-order relations between properties and they are necessary insofar as they hold in all possible worlds in which the relevant first-order properties exist.

"Laws of nature – a sceptical view" is presented by Uwe Meixner. After examining van Fraassen's sceptical argument as well as the so-called TAD-approach (Tooley/Armstrong/Dretske), Meixner explains his own idea. He sees a way out of the sceptical dilemma, i.e. the tension between laws of nature transcending the phenomena, on the one hand, and keeping these laws within our epistemic reach, on the other hand. The solution, he thinks, can be achieved by 'relativising' laws of nature to our beliefs and decisions. These laws are made by us and they can be rescinded by us.

There is an obvious connection between new scientific discoveries, new laws and new properties. To understand laws we need an understanding of properties, and this is discussed in Johannes Persson's "The law's properties". He presents a framework for distinguishing between properties and fake properties that allows us to handle questions concerning the ontological status of laws. Furthermore, this approach proves suitable in discussing tests for properties suggested by Maxwell, Ramsey and Cartwright.

Gerhard Schurz considers "Laws of nature versus system laws". The first are fundamental laws of physics that hold everywhere in the universe. System laws concern some specific systems in a given time under concrete specification of all forces acting within or upon the system. Schurz draws a three-fold distinction which recognises that: 1) *ceteris paribus* clauses are needed for system laws, not for laws of nature; 2) there are universal conditionals functioning as system laws; and 3) most system laws are fundamental (i.e. non-derived), whereas some laws of nature are derived.

The volume ends with Werner Stelzner's paper on "Psychologism, universality and the use of logic". Here we find another perspective on the overall theme of the book. Assuming both rules of nature and of language, Stelzner asks what is specific about rules of logic? Rules of logic are treated as special language rules that span the realm for explicating and systematising rules of nature. In order to make this clear, he paints a very detailed picture of the rise of modern logic between psychologism and antipsychologism.

References

D. Armstrong. *Universals and Scientific Realism*, 2 volumes. Cambridge University Press, Cambridge, 1978.

D. Armstrong. *What is a Law of Nature?* Cambridge University Press, Cambridge, 1983.

H. B. Callen. *Thermodynamics and an Introduction to Thermostatistics.* John Wiley, New York, 1985.

R. Carnap. *The Philosophical Foundations of Physics.* Basic Books, New York, 1966.

N. Cartwright. *How the Law of Physics Lie.* Clarendon, Oxford, 1983.

N. Cartwright. *Nature's Capacities and their Measurement.* Clarendon, Oxford, 1989.

N. Cartwright. *The Dappled World.* Cambridge University Press, Cambridge, 2000.

D. Davidson. *Essays on Actions and Events*, "Actions, Reasons and Causes", pages 3–19. Oxford, 1980.

D. Davidson. *Essays on Actions and Events*, "Causal Relations". Oxford, 1980.

F. Dretske. Laws of nature. *Philosophy of Science*, 44:248–268, 1977.

P. Duhem. Une nouvelle théorie du monde inorganique. *Revue des questions scientifiques*, 33:99–133, 1893.

P. Duhem. *The Aim and Structure of Physical Theory.* Princeton University Press, Princeton, 1991.

N. Goodman. The problem of counterfactual conditionals. *Journal of Philosophy*, 44:113–128, 1947.

N. Goodman. Review of Reichenbach (1947). *Philosophical Review*, 58:100–102, 1948.

N. R. Hanson. *Patterns of Discovery.* Cambridge University Press, Cambridge, 1958.

D. Hume. *A Treatise of Human Nature.* Clarendon Press, Oxford, 1958.

C. G. Hempel. Studies in the logic of confirmation. *Mind*, 54, 1945. reprinted in *Aspects of Scientific Explanation*, Free Press, Collier-Macmillan, Toronto, 1965.

C. G. Hempel. Review of Reichenbach (1954). *Journal of Symbolic Logic*, 10:54–55, 1955.

Hempel, C. G. (1965). *Aspects of Scientific Explanation.* The Free Press, New York.

Hempel, C. G. and Oppenheim, P. (1948). Studies in the logic of explanation. *Philosophy of Science*, 15:135–175. (Reprinted in Hempel, 1965: 245–290).

E. K. Jobe. Some recent work on the concept of law. *Philosophy of Science*, 34:363–381, 1967.

W. A. Kneale. Natural laws and contrary-to-fact conditionals. *Analysis*, 10, 1950.

W. A. Kneale. Universality and necessity. *The British Journal of Philosophy of Science*, 12, 1961.

T. B. Kuhn. *The Structure of the Scientific Revolutions.* The University of Chicago Press, Chicago, 1962/70.

G. W. Leibniz. *Leibniz: Selections*, "The Metaphysical Foundations of Mathematics". Charles Scribner's Sons, New York, 1951.

D. Lewis. *Counterfactuals.* Basil Blackwell, Oxford, 1973.

D. Lewis. *On the Plurality of Worlds.* Basil Blackwell, Oxford, 1986.

D. Lewis. *Philosophical Papers.* Lewis, 1986b, chapter Counterfactual Dependence and Time's Arrow (incl. Postscripts), pages 32–66.

D. Lewis. *Philosophical Papers*, volume II. Oxford University Press, Oxford, 1986.

E. Nagel. *The Structure of Science.* Routledge & Kegan Paul, London, 1961.

P. Needham. Duhem's physicalism. *Studies in History and Philosophy of Science*, 29:33–62, 1998.

P. Needham. Duhem and Quine. *Dialectica*, 54:109–132, 2000.

A. Pap. Disposition concepts and extensional logic. In M. Scriven H. Feigl and G. Maxwell, editors, *Minnesota Studies in the Philosophy of Science*, volume 2, Minnesota, 1958. University of Minnesota Press.

H. Poincaré. *Science and Hypothesis.* Dover, New York, 1952.

K. R. Popper. *The Logic of Scientific Discovery.* Hutchinson, London, 1959.

F. P. Ramsey. *Foundations: Essays in Philosophy, Logic, Mathematics and Economics.* Routledge & Kegan Paul, London, 1978.

T. Read. *The Works of Thomas Read*, "Essays on the Intellectual Powers of Man". Ed. by Sir W. Hamilton. Edinburgh, 7th edition, 1872.

H. Reichenbach. *Elements of Symbolic Logic.* Macmillan, London, 1947.

H. Reichenbach. *Nomological Statements and Admissible Operations.* North Holland, Berkeley and Los Angeles, 1954.

Reissued under the title *Law, Modalities and Counterfactuals*, University of California Press.

H. Reichenbach. *The Philosophy of Space and Time*. Dover, New York, 1957. trans. of *Philosophie der Raum-Zeit-Lehre*, (Berlin, 1928).

B. Stroud. *Hume*. Routledge & Kegan Paul, London, 1977.

M. Tooley. The nature of law. *Canadian Journal of Philosophy*, 7:667–698, 1977.

B. C. van Fraassen. *The Scientific Image*. Clarendon Press, Oxford, 1980.

B. C. van Fraassen. *Laws and Symmetry*. Clarendon Press, Oxford, 1989.

WHY ARE (MOST) LAWS OF NATURE MATHEMATICAL?

Mauro Dorato

Our experience hitherto justifies us in believing that nature is the realization of the simplest conceivable mathematical ideas

—Einstein

In a frequently quoted but scarcely read paper, the Hungarian physicist Eugene Wigner rediscovered a question that had been implicitly posed for the first time by the *Transcendental Aesthetics* of the "Critique of Pure Reason". More precisely, rather than asking, in the typical style of Kant, "how is mathematics possible", Wigner was wondering how it could be so "unreasonably effective in the natural sciences" (Wigner, 1967).

The effectiveness in question refers to the numerous cases of mathematical theories, often developed without regard to their possible applications, that later have played an important and unexpected descriptive, explanatory and predictive role in physics and other natural sciences. A frequently given example is that of the conic sections, already known by the Greeks before Christ and used by Kepler many centuries after their discovery to describe the orbits of celestial bodies. Even more striking is the case of non-Euclidean geometries, applied by Einstein to describe how heavy matter bends the structure of spacetime in his general theory of relativity: the theory of curved, non-Euclidean spaces had already been built a century earlier by Gauss, Lobacevski and Riemann. A literary quotation addressing the role of complex numbers, due to the German writer Robert Musil, will conclude my necessarily short list of

J. Faye, P. Needham, U. Scheffler and M. Urchs, (eds.), Nature's Principles, 55–75.

examples: "The strange fact is that with these imaginary or even impossible numbers one can anyway make perfectly real calculations which end in a concrete result". Ironically, at the time of *The Confusions of the Young Törless* (1906), from which this passage is taken, Musil could not be aware at the fact that the most successful theory of the atomic structure of matter – quantum mechanics – would have been using imaginary numbers to calculate the probability of measurements.

In this paper, I want to raise once again Wigner's question (to which I will be referring as 'WQ') *in order to shed light on the related issue of the nature of scientific laws.* Namely, my main purpose is to show that typical questions of the philosophical literature on laws, like

1 what laws are[1] and

2 how we come to know them,

can be fruitfully approached afresh if we pay due attention to *their mathematical character.* Note first of all that if we replied to WQ by saying that "nature itself is mathematical", we would trivialize the question only in appearance, since such a metaphysical answer should itself be explained: if, say, mathematics is a creation of ours, why are laws of nature itself mathematical? On the other hand, the fact that the laws of science are mathematical poses the question of the mathematizability of nature, which provides the clue for a correct understanding of WQ.

Considering that the idealized and simplified character of physical laws – on which many philosophers have insisted – might be simply due to the fact that such laws are mathematical and mathematical equations must be *tractable and amenable to solutions*, it is quite surprising that no one has tackled the philosophical problems of scientific laws from the perspective of Wigner's question. To be sure, a possible explanation of this neglect could come from the immediate remark that *not all laws are mathematical or expressible in a quantitative language* ("metals tend to expand if heated" or "all ravens are black" are but two examples). However, most if not all physical laws are expressed mathematically, and mathematical models have become increasingly important in biology (think of evolutionary game theory), in the cognitive domain (neural networks), and in the social sciences (decision theory and game theory).

[1]Recall that, roughly, ontological realists about laws claim that statements expressing laws of nature refer to mind-independent properties and relations obtaining between natural systems, while semantic realists claim that law statements are approximately true, or at least are susceptible of receiving a definite truth-value. Ontological antirealists about laws deny them any referential character, while semantic antirealists regard law-statement as merely useful, but do not grant them the property of being true or false.

This fact justify an approach to laws that focuses on attempts at answering WQ.

In the first section of the paper (1), I will illustrate the significance of WQ for the issue of natural laws, as well as for the philosophy of mathematics and language, by putting the problem in context and by trying to argue that the problems it raises are *genuine*. This preliminary task is necessary, since if WQ were a pseudo-problem, the claim that it could help us to consider the issue of scientific laws in a new way would be groundless. In the second and third sections (2-3), I will present an important, current attempt at answering WQ, centered on the view that laws of nature are the *software* of the physical universe. In this metaphor, which for its proponents is suggestive of a deeper truth about laws, the universe is considered to be a gigantic computer whose *hardware* is whatever fundamental physics tells us about the ultimate component of matter (fields, particles, superstrings, etc) and whose software is the ordered sequence of states it goes through in time. In the fourth and fifth sections (4-5), I will critically discuss this approach to natural laws by raising various difficulties, some of which appear to be fatal. Finally (6), I will propose my own way to relate WQ to the issue of the nature of scientific laws, which recommends to look more closely at the complex operations of *measurement.* While as a solution to WQ this proposal is only a suggestion to look in a new direction, it helps us to see why mathematical laws represent the world in the same way (or very analogously to the way) in which a quantitative order introduced with a scale represent the relations between the properties of the systems to which we apply it.

1. Is Wigner's problem genuine? A puzzle in the philosophy of math and language

First of all, let us begin with the skeptical or even cynical remark that for every philosopher that regards a problem as fundamental and ineluctable, there is at least another one that claims that the same problem is not genuine. In our case, someone that might want to diminish or even dissolve the sense of mystery of the unexpected applicability of various parts of abstract mathematics to the description of the natural world, might want to point out that, first of all, *important branches of mathematics were explicitly developed to solve physical problems.* The case of *calculus* in Newton's *Principia* is an important example of a piece of mathematics that can be used to describe the relationship between certain physical magnitudes simply because it was devised for that purpose. Secondly, one could remark that *many parts of mathematics have no application whatsoever*, and Wigner's problem could simply be the effect of selection: for few pieces of mathematics that are applicable many more simply aren't, but we obviously take notice only of the former and not of the latter.

As a reply to the first objection above, consider that even if the motivation to create calculus or the theory of probability has come from empirical problems (physical questions of instantaneous speed or "less intellectual", combinatorial questions originated with dice tossing, respectively), this historical fact does not diminish at all the sense of mystery created by WQ. Let us suppose that we can divide all successful applications of mathematics into two classes, the expected and the unexpected ones. Though the latter make a greater impression upon us, and are therefore favored in the illustration of WQ for obvious rhetorical reasons, it is plausible to claim that also the former, sooner or later, will generate the same type of wonder of the unexpected ones.

For instance, note that once launched on its path after the applicative input, calculus *proceeded independently of empirical motivations or intended applications.* If it must be acknowledged that *advanced* calculus has applications to physics that had not been intended at all by the mathematicians that built on Newton and Leibniz, it then remains true that mathematical theories "give us in return much more than we originally put into them".

As a way to illustrate what the latter metaphor mean, consider that the relationship between mathematics and the empirical world resemble in this respect the relationship between the *theoretical* and the *observational* terms of a theory, as the mature Hempel understood it (Hempel, 1958). Mathematical theories, exactly like the theoretical terms of a theory, have an *open texture*, in the sense that the range of applicability of the most fruitful of them cannot be exhausted by the original, intended application for which they had been devised. In the same sense, the original operational definitions with which some theoretical term has been introduced does not give a translation of the latter, if the term is really fruitful.

In a word, the heart of the problem of the applicability, it seems, is that mathematics proceeds with a demonstrative method that is *a priori*. Physics and the other natural sciences are instead based on experiments, and the representative component of physical laws or models is certainly derived from our observations, and is therefore *a posteriori*. How is it possible that mathematical theorems, arrived at *via* rigorous, *a priori* derivations, can help us to discover properties of the natural world, which we can know only with the help of our observations, and therefore *a posteriori*? I think that this way of formulating Wigner's question –

which possibly was in the back of Kant's mind when he thought of mathematics as based on *synthetic a priori judgments* – clearly disposes of the skeptic's first objection.

As to the second one, based on the selective character of the examples usually given to illustrate WQ, consider that even if the case of the non-Euclidean geometries were the *only one* in favor of the significance of WQ, it would be so striking as to deserve an explanation. The fact is that it is difficult to think of a major area of mathematics that is really "immune" from applications (besides differential geometry, think of algebra and algebraic topology for particle physics). Furthermore, "pure" mathematicians trying to insulate their discipline from possible applications are either disappointed by some unexpected, later match with the empirical world, or do not serve their discipline in the best possible way.[2]

Taking now for granted that WQ is genuine, let us note in passing that it should be discussed much more often than it is also in order to evaluate various philosophical positions on the nature of mathematical knowledge. *Prima facie*, it creates problem to many of such positions, and it is perhaps for this reason that it is so neglected.

Platonism is notoriously affected by the problem of explaining how it is that we can know, and therefore *causally interact with, a world of abstract entities which we discover*, and that has therefore not been created by our language or minds. Remembering that *abstract entities are causally inert by definition*, this objection clearly holds, provided that knowing x presupposes to some extent that we must causally interact with x. However, even if we could circumvent this problem by denying the premise of the causal theory of knowledge, we would have to explain why the *physical* world, which is spatio-temporally extended, should mirror the structure of the *mathematical* world, in such a way as to allow the application of concepts inhabiting the latter to the former.

Within *constructivist* positions, we must ask why a *creation* of ours can carry so much descriptive and predictive power, enabling us to explain and systematize entities of the natural world which we obviously did *not* create. *Prima facie, it would seem that unless mathematical notions derive from our experience, it is difficult to make sense of the applicability of mathematics.* However, even if certain mathematical structures were the evolutionary product of a long process of cognitive adap-

[2]As von Neumann once wrote, separating mathematics from its empirical applications can only transform it into a sterile intellectual game. This of course does not mean that mathematicians should not continue to pursue their work without having some application in mind, since it is impossible to predict which mathematical branch will be susceptible of being more fruitful.

tation of our brain to objects whose size is comparable to our bodies, we would have to explain why these structure have been successful also in describing entities whose size is much smaller and much bigger than our bodies by various orders of magnitudes (think of the application of group theory in particle physics).

One further attempt at explaining WQ that might sound slightly deflationary comes from the claim that mathematics are "effective" for the same reasons why our natural languages are.[3] After all, mathematics is a particular type of language, only more abstract and semantically more precise than ordinary languages, and its applicability to the external world should generate no more surprise than the fact that we use Italian, English or Danish to refer to the world around us (we "apply" them).

Furthermore, since *mathematical abilities* are to a good extent genetically determined, fundamental mathematical concepts, like *number* or *space*, might be innate, in the same sense in which fundamental concepts are innate in Fodor's *language of thought*: otherwise, we might ask, what would be the *object* of such mathematical abilities? If we suppose, in addition, that the contents of our thoughts are expressed in symbolic structures of an innate language, whose syntax and semantics are similar to (though more abstract than) those of the natural language, *then* the claim that all our mental processes are *computational* would explain the *deductive* character of mathematics and would also explain why mathematical knowledge is *a priori* without invoking any outlandish form of Platonism.

The connection between WQ and the philosophy of language is undeniable, and would deserve much more empirical and conceptual attention than it usually does. However, here I feel justified in simply mentioning it, since I surmise that a further exploration at this stage of our knowledge of computational knowledge would not help us to come any nearer an understanding of the nature of scientific laws.

In any case, also the prospect of a linguistic approach *vis à vis* the applicability question might raise some reservations. For one thing, the mathematical method of proving, that is, of ascertaining the justifiability of an assertion *via* deductions, is unparalleled in "natural" uses of natural languages. Of course, we can use the rules of logic to deduce an assertion from other assertions, all cast in a natural language. However, this reply has little force, to the extent that the use in question is already part of a regimented way of conducing our intellect, *logic*, which is itself

[3] I owe this suggestion to Jan Faye, with whom I have had several conversations on this problem.

a branch of mathematics: in this sense the reply is part of the problem we are trying to clarify, which is why mathematics is effective. What is more important, the *epistemology* of natural language and of mathematics *seem* very different, insofar as the latter makes use of *a priori* knowledge in a much more systematic and extended way, while within the former the (referential and causal) contact with the empirical world apparently plays a much more important role (*a posteriori* knowledge).

2. Three ingredients to make mathematical laws

Our species is characterized first and above all by its ability to build artifacts. As a consequence, in the attempt to try to understand the unfamiliar and the unknown in terms of the known and the familiar, *the latter has often been equated with what we can build, for the simple reason that we know how it works.* Given the immense importance that computers have in our society, it can come scarcely as a surprise that physicists and philosophers have relied on computer science not only in order to understand the nature of laws, but also in trying to give a tentative answer to Wigner's question about the applicability of mathematics.

Essentially, the idea is that we can look at any physical system from two viewpoints, an "anatomic" one, which means that we look at its components (corresponding to the *hardware* of a computer) and a physiological, functional one, which means that we look at what the system does, or the laws controlling its behavior (corresponding to the software of a computer). If the universe can be regarded as one unique, gigantic physical system, the laws that govern its unfolding in time could be regarded as *the software of the universe.* In order to further explore this metaphor and understand its significance for the question of laws, consider that, following Whewell, there are three components that are necessary to express a law of nature in mathematical language:

1 the initial or boundary conditions, i.e., *the inputs*, or the numerical sequence that we obtain *via* a measurement;

2 the *algorithmic structure*, given by the mathematical formula that we apply to the data in 1;

3 the quantities that are left invariant by the application of 2 to 1, namely the *constants of nature.*[4]

[4]For this tripartite structure, see also Barrow, 1988, p. 279.

Whewell referred to 1 as the determination of the independent variable; to 2 as the discovery of the formula that is capable of "colligating" the independent to the dependent variable ("the colligation of facts"), and to 3 as the determination of the coefficients (see Butts, 1968, pp. 210, 211).

I think it is safe to claim that a philosophical discussion on the nature of scientific laws that did not take these factors into account would not be complete, and would run the risk of separating the philosophical analysis from the scientific practice and from what scientists usually mean by "scientific law". Even though not all scientific laws are mathematical, it is not implausible to suppose that many neo-positivist accounts of laws suffered from the original sin of supposing that universal, but "*qualitative*" statements of the kind "all ravens are black", could be considered paradigmatic examples of natural laws.[5] Such statements do not contain any of the three ingredients mentioned above, and it is perhaps for this reason that a good part of the philosophical debate on laws – think of the failed attempts to separate genuine laws from universal generalizations on a purely syntactic level – has often been so remote from the scientific practice to become a purely academic game.

I do not think that I am endorsing here a form of mathematical chauvinism. Clearly, since the mathematization of laws is a phenomenon that prevails in particular within physics and some branches of biology and economics, it is clear that it is in these empirical sciences that the problem of why laws are mathematical apply. However, it is not difficult to show that what is philosophically relevant about the way in which mathematically expressed law represent the world can be extended without too many difficulties to the laws of other, less mathematized sciences. By quoting Kant, and thereby using for once the principle of authority, we can consider the more mathematized sciences to be a model of objective knowledge from which the other sciences draw an inspiration: "Since in any theory about nature one can find so much science, properly speaking, as there is knowledge *a priori*, it follows that the doctrine of nature can contain so much science, properly speaking, as there is mathematics that can be applied to it." (Kant, 1968, p. 470 – my translation)

3. Laws as the software of physical systems

Focusing just on the first two components of the three points presented above (the third is not relevant for our purposes), we can regard

[5]Here "qualitative" refers to the purely classificatory nature of the predicates (in the example, black) appearing in the universally quantified statements.

a mathematically formulated law as a "bridge" colligating two banks of a river, each constituted by quantitative data resulting from measurements. On one side of the river we find the initial data or boundary conditions – 1 above – which in our metaphor we can regard as the input, and on the other side we find the predictions or retrodictions – the *output* – the result of a *calculus.*

Since such a result is obtained in a purely deductive fashion, that is, thanks to the application of the law to the initial data, the metaphor of the scientific laws regarded as the algorithm of a computer appears initially justified. If the initial data in fact are such as to satisfy some mathematical conditions which in this context can be omitted,[6] and whenever the solution to the equation exists and is unique, a mathematical law expressed as a differential equation enables us to transform in a finite numbers of steps, and in purely mechanical fashion, the initial data in final predictions (output).

What interests us is, of course, whether such an analogy between the laws of succession of any physical system – regarded as something that evolves in time by going through a finite number of states describable in physical language – and the software of a computer, can help us to better understand: (i) why the world is describable by mathematical laws and (ii) how the latter are related to the world, that is, how they *represent* it. In order to shed light on the presuppositions of the law-software analogy, we should ask whether also a physical system, in a sense to be specified, can be said to "compute" its "next" state *by causing it.* Can we say that a physical system going from an initial to a final state *literally executes a program or calculates its future state*, in the same sense in which a mathematical physicist deduces or calculates the predictions corresponding to the initial data?

On the basis of a physical version of Turing-Church hypothesis (according to which every intuitively computable function can be computed by a Turing machine), David Deutsch has recently tried to answer these questions by claiming that every physical process that can be performed in a finite number of steps can be *simulated* by a *quantum Turing machine* ("quantum" is needed because energy levels of finite systems must be, unlike classical ones, *discrete*) (Deutsch, 1985). If Deutsch's hypothesis proved to be correct, the relationship between a physical system undergoing a finite number of successive stages and a quantum Turing machine would be one of *simulation.* In this respect, the analogy between any physical system undergoing a temporal evolution in finite time and

[6]The functions representing the data must be differentiable at least as many times as the degree of the differential equation giving the algorithm.

the mathematical physicist performing a calculation would be given by the Turing machine acting as a "mediator": the capacities of both a natural system an of a human calculator (the physicist) to perform a series of operations in succession can both be simulated by a Turing machine, *if* the Turing machine hypothesis about the human mind holds water.

Suppose that the temporal evolution of any physical system is describable by *finite strings of real numbers*, corresponding to operational measures of physical magnitudes (temperature, pressure etc.). We can have two cases: such string can be ordered

$$(111000111000111000\ldots)$$

or truly random

$$(0100110101100110\ldots)$$

In the first case, the string can be generated by a simple instruction ("print 111000 n times"), which is much shorter than the list itself. In the second case, the string *appears* as truly random, where "appear" is meant to stress that while we can show that a finite string is not random by giving the generating law (algorithm), we can never prove that a string is random because of rigorous results in classical logic (the halting problem).

At this point we can give two definitions, based on algorithmic complexity theory, which will be relevant for our purpose:

DEF 1 *the complexity of a string is the length of the shortest algorithm capable of generating it,*

DEF 2 *a string is algorithmically compressible when there is an algorithm capable of generating it, such that its information content (number of bits) is much less than that of the string.*

As an illustration of these definitions, consider that a string like

$$\{1, 4, 9, 16, 25, 36, \ldots\} \tag{1}$$

is obviously not random, since it can be obtained by squaring the positive integers in the list

$$\{1, 2, 3, 4, 5, 6, \ldots\} \tag{2}$$

If the numbers in 2 correspond to measured magnitudes in such a way that, say in a temporal interval of 1, 2, 3 seconds (the input data), a body travels 1, 4, 9 meters (the output), then the existence of a rule generating 1 from 2 shows that 1 is *algorithmically compressible*. The above algorithm is – *modulo* the constant $1/2g$. – Galileo's law of free fall, generating the spatial intervals 1 from the square of the temporal intervals 2.

In a word, by following the metaphor of scientific laws regarded as the software of a physical system, we discover that *searching for laws is tantamount to asking which is the length of the shortest program capable of generating the string of numbers expressing the experimental measures.* Such a length – the complexity of the string – will be equal to that of the original string only if the latter is composed by apparently random numbers, and does not obey any known law.

The idea that scientific laws are an economic synthesis of all the information contained in our observations is certainly not new, and in this algorithmic approach it finds a new, rigorous and precise formulation. It was especially Ernst Mach who regarded science and its theories and laws as a summary of our observations. As he wrote: "Science is a form of business. Its purpose is to find the maximum amount of the infinite eternal truth with the minimum amount of work, in the minimum expenditure of time and with the minimum amount of thought effort".

After having made explicit the philosophical consequences that seem to follow from the software metaphor for scientific laws, we can now finally discuss a possible explanation of the applicability of mathematics, due to the physicist John Barrow: "science exists because the natural world seems algorithmically compressible. The mathematical formulae that we call laws of nature are economical reductions of enormous sequences of data expressing changes of state of the world: here is what we mean by intelligibility of the world ... *Since the physical world is algorithmically compressible*, mathematics is useful to describe it because it is the language of the abbreviation of sequences. The human mind enables us to make contact with that world because our brain has the ability of compressing complex sequences of sense data in shorter form. Such abbreviations make thought and memory possible. The natural limits that nature poses to our senses prevent us from overloading our brains with information about the world. Such limits are security gates for our minds" (Barrow, 1992, pp. 93–96).

4. Does the software metaphor really work?

Let us now go into discussing the various difficulties of this interesting proposal, both as an answer to WQ and as a response to the issue of natural laws. If laws are nothing but *compressed observational information*, let us remark at the outset that the focus of this approach on laws is clearly *epistemic.* This seems to be confirmed by what Barrow claims about the brain, regarded as the main filter of information coming from sense data.

Unfortunately, claiming that laws are merely a more concise rendering of observational information seems to trap this view into the Procrustean bed of the early neo-positivist conception of theories, and of theoretical terms in particular. First of all, how can we explain the fact that known scientific laws can be often used to explain and predict wholly new phenomena if laws are – exactly like the theoretical terms were in the early neo-positivistic conception – a mere translation, that is, a shorter, more compact version, of observational statements entailed by measurements? Secondly, how can we account for the fact that some laws are purely theoretical and, as such, make no reference whatsoever to directly observable phenomena like phenomenological laws do?

Finally, how can we make sense of the *abstraction* and the *idealization* implicit in the construction of phenomenological and theoretical laws within the algorithmic conception of laws? In the actual world we certainly observe frictional effects, but the law of free fall and that of the isocrony of the pendulum *abstract* from them. From this viewpoint, *rather than passively reflect and summarize our observations*, physical laws select and "caricaturize" the property of a physical system.

Perhaps the defenders of the algorithmic approach to scientific laws can reply to each of these objections by pointing out that the *form* that a law takes in one field may have an heuristic role. For instance, it may suggest interesting *syntactical analogies* with the relations exemplified by data coming from unrelated fields, in the same sense in which Coulomb's law in electrostatics is syntactically related to Newton's law in mechanics. Even supposing that formal analogies of this sort are able to explain the *open character* and the fruitfulness of good theoretical laws, i.e., their being not confined to the data for which they had been devised, another question remains unanswered: why do formal analogies work? To this question the software approach to laws does not seem to be able to provide any satisfactory answer, except the one that consists in pointing to the repeatability of some patterns in the world as a brute fact of nature. On the epistemic side, it is easier to point out that we try to make the most of the symbols we have already successfully applied, and the more observations fall under a given mathematical structure, the simpler, the more economical, and more unified is our explanation of the world.

To come now to the second objection, involving theoretical laws that do not mention any directly observational term, we can always decide to make the observational-theoretical distinction a matter of degree, or to abandon the dubious notion of "direct observation". Don't physicists claim to observe, albeit indirectly, atoms and electrons? In this way, one can maintain that it is not just phenomenological laws that can be captured by the software metaphor, but also those theoretical laws regulating the behavior of entities that are only indirectly observable (atoms, electrons etc.).

Finally, in order to give due attention to the constructive, and selective character of physical laws (the third objection) it can be observed that the abstraction and idealization typical of physical laws might take place *before* their formulation, thanks to *a judicious choice of which quantities must be measured in order to find a formula colligating them.* This reply seems objectively the weakest of all, given that the law of inertia for example does not seem to depend on any choice of which quantities to measure.

The difficulties this position has in explaining WQ are even more serious, since it not clear what the relationship should be between the above mentioned capacity of the brain of filtering information in perceiving the world and the applicability of mathematics, even when the latter is regarded as the art of compressing sequences. Even the more ontological suggestion of Barrow's quotation does not help much, at least until we have a better grasp of what it means to claim that "the physical world is algorithmically compressible". Isn't this another way of formulating what we are trying to explain, namely the link between the regularity and the repeatability of certain phenomena of the world and our use of mathematics? An attribution "to the physical world in itself" of the compressibility in question seems to imply an acceptance of a classical empiricist, regularist position, according to which the objective content of any law is simply a spatiotemporally valid regularity. It is the repeatability of the phenomena subsumed under a law that is responsible for "the *redundancy* of the world", enabling us to compress the information coded in our observations.

A further difficulty of the information-theoretic approach to laws lies in the way in which the view is formulated: if it does not make sense to claim that a physical system literally and really computes, a "Turing test" for physical systems (involving a physical system regarded as a black box and a Turing machine observationally indistinguishable from it) becomes meaningless.

However, in response to this objection, consider that no physical systems, not even a computer, really "calculates" or "computes", if by such terms we refer to an *intentional, conscious act* accompanying a goal-oriented activity. Computers executing a program just undergo physical changes of states that *we interpret* as a computation, on the basis of a task that we have the machine perform for us. Certainly, physical laws must be such as to enable us to use physical processes to perform addition and multiplication, but any physical process, like our heartbeat or the motion of the Earth, can be used as a measuring and therefore as a calculating device. Clearly, such a use presupposes an intentional act of attribution of a function.

In this regard, to claim that a natural system *computes* its future state is no more metaphorical than saying that a computer calculates, given that both statements presuppose a function that we attribute to an inanimate object. Consequently, it is not possible to attack the identification between natural laws and algorithms executed by computers on the basis of the fact that only the latter *literally compute.* In fact, *either physical systems and computers both compute or they both don't.* And since we are inclined to call the machine I am using right now to write "computer", there is a sense in which we can extend this label also to more general physical systems.

5. Two fatal objections to the information-theoretic approach to laws of nature

If the objections we had examined so far can perhaps be rebutted, the two that I am about to present seem fatal to the whole algorithmic view of laws. The first simply points out that not all physical laws relate states ordered by the relationship of temporal succession "later than" (*laws of succession*). Many of them constrain physical states S existing *at the same time*, in such a way that no two such S's can be connected by causal or luminous signals (*laws of coexistence*).

The trouble might have already been anticipated by the reader. (i) If laws of coexistence link *spacelike-related* properties of physical systems; (ii) if *the notion of algorithm is essential sequential and temporal* (even in parallel computations, the results of distinct calculations must interact before the output); (iii) if laws refer to something existing mind-independently, then from (i) (ii) and (iii) it follows that *either all laws of coexistence can be reduced to laws of succession, or natural laws in general cannot be assimilated to computations or algorithms.*

Consider laws of coexistence like Newton's law of gravitational attraction, Gauss' law relating the electric flux through a close surface Φ generated by an electric charge q located within the surface, and Boyle's law relating pressure, volume and temperature:

$$F = G(M_1 M_2)/r^2$$

$$\Phi = \varepsilon_0 q$$

$$PV = kT$$

Except the first, which is an approximation to the more fundamental field equation of general relativity, these laws don't assume instantaneous action at a distance, so they are not in conflict with relativity or field theories, and neither can they be *prima facie* reduced to causal laws. Moreover, *despite the fact that they enable us to deduce or calculate one magnitude from the others* (say F or Φ or PV can be calculated from the magnitudes on the right-hand side of the equality sign of the equations), we cannot interpret them in an ontic way, namely *as if the system could really calculate one state from the one that is related to it in the mathematical formula.* In a word, the sequential character of laws is a necessary requirement for the view that laws are algorithms, given that the latter notion is intrinsically temporal: how could a Turing machine simulate the behavior of a physical system that is described by laws that merely specify the correlation of values in a spacelike hypersurface?

Clearly, if one wants to defend Deutsch's thesis, one must claim that all laws of coexistence are somehow less fundamental than laws of succession and depend (supervene) on the latter in some sense. Can we claim that all laws of coexistence *supervene* on laws of succession, in the sense that the origin of a law of coexistence presupposes a law o succession in the same sense in which a correlation between spacelike-related events *normally* presupposes a common cause in their past? If this supervenience thesis proved correct, the main obstacle against an attempt at giving an ontological interpretation of the algorithmic view of laws would be removed, because the existence of a law of coexistence would always presuppose a law of succession.

Unfortunately for this suggestion, any attempt at showing that laws of succession are more fundamental than laws of coexistence in the sense given above must face the existence or quantum correlations at a superluminal distance. For many philosophers, such correlations exclude the possibility to invoke a common cause in the past to explain the spacelike-related correlations among measurement results.[7] However, following Nancy Cartwright, we could regard the interaction of the two particles in the past *as a probabilistic common cause*, despite the lack of screening-off feature or of factorizability (Cartwright, 1994). But even if we managed to show that all quantum laws of coexistence originated from common causes and therefore from causal laws, this would not be equivalent to say that the laws of coexistence don't exist. And their mere existence is sufficient to cast serious doubts on the whole information-

[7] See for example van Fraassen, 1982.

theoretic approach to laws, unless, of course, its defenders were content with a mere epistemic interpretation. While there is nothing wrong in principle in limiting the impact of the algorithmic view to the epistemic side of the debate on laws, this self-imposed restriction seem to deplete the view of its significance. The algorithmic view of laws is interesting when it carries ontic implications: if it is just a way of rephrasing the well-known fact that we use differential equations to calculate the outcome of our measurements one might as well abandon it.

The second difficulty of the algorithmic view is the existence of non-computable equations in current physical theories, something which would clearly give a fatal blow to the association of any physical system to a Turing machine. In general, the computability of a dynamic equation and of initial data does not guarantee the computability of the solution in three distinct cases: when the solutions are not unique, when they are obtained from unbound operators, and when the function representing the solution is neither differentiable nor continuous and is therefore "weak" in the sense of the theory of distributions.[8]

In a word, the software approach to scientific laws does not appear to be sufficiently general to cover the whole range of cases in which laws expressed as differential equations are used in physics. As such, and despite its interest in shedding light on *some* important aspects of scientific laws (their being an economic compression of observational information), it cannot be regarded as an acceptable solution to Wigner's problem and to the problem of explaining in which way laws refer to the world.

6. Measurement as a key to WQ and to understanding nature of laws

Before advancing my own proposal to the problem, let me warn the reader that its appropriate development and defense is the topic of another essay.[9] Here, I can only outline the guidelines of a view that regards measurement not just as the main business of science, but as the key to give a convincing answer to WQ and to the problem of whether and how mathematical laws represent the world.

Let us start by advancing the plausible claim that a *quantitative* treatment of phenomena (or what Suppes calls "data modeling") is a necessary condition of both measurement operations and the applicability of

[8]For a thorough treatment, see Pour-El and Richards, 1989.

[9]See the second chapter of my forthcoming *The Software of the Universe. An Introduction to the History and Philosophy of Laws of Nature*, Ashgate.

mathematics. We should now ask what kind of relationship there might be between the *qualitative* fabric of phenomena – what appears to our senses – and their *quantitative* treatment. Here are some points that I will motivate only schematically, since I regard them as very plausible.

1 Laws *relate* properties P of entities or natural systems. In the examples above, the gravitational force between two bodies is related to their masses and distance, and the pressure and volume of an ideal gas are related to its temperature. *Briefly put, laws are relations between properties P of a system S.*

2 Such properties can become quantitative – scalar or vectorial as they may be, like "having a mass of a certain magnitude" or "having a certain velocity" – only after having introduced a metric (a scale) and having performed some measurements.

3 Since the attribution of particular (real or rational) numbers to properties of events is scale-dependent and conventional, *it is only relations among physical entities that are preserved by their mathematical models.* If I say that "today's temperature is twice as high as yesterday's", my statement is not objective, or rather, has no definite truth value, unless I specify to which measuring system I am referring it (Celsius, Fahrenheit or Kelvin). In any scale, however, what is preserved of the above statement is clearly that today's temperature is *hotter than* yesterday.

4 By recalling Carnap's good old distinction among classificatory, comparative and quantitative concepts,[10] we can regard the distinction between qualitative and quantitative language as a conventional distinction to which nothing in the real world corresponds. This position is tantamount to claiming that the mathematization of the world preserves only some *relevant relations* about the represented natural systems.

5 The key notion to getting close to an understanding of WQ is that of *a partial isomorphic relation between (parts of) a mathematical model and the represented phenomena*, something which entails the existence of a *structural resemblance* between them. Isomorphisms are in fact relations-preserving bijective maps between models and represented phenomena.[11]

---

[10] See Carnap, 1950.

[11] For an analogous view, see Bueno et al., 2003.

As an illustration of the previous point, suppose we have transitive, asymmetric relations $<_p$ or $<_l$ holding, respectively, among differently plausible versions of witness reports ($<_p$) , or among objects of different length ($<_l$). If we write '$W_1 <_p W_2$' we mean that the report of the witness W_1 is less plausible than the report of the witness W_2, while '$O_1 <_l O_2$' means that O_1 is shorter than O_2. By introducing comparative concepts we are able to refine, order and compare our *intuitive, primitive classification* of, respectively, stories as being "plausible" and "implausible" and objects as being or "short" or "long". We also need equivalence relations "$=_p$" "$=_l$", which partition all our stories and all our objects into disjoint classes, whose members having "same plausibility" and "same length" respectively, whatever our methods to establish such judgements.

The representability in numerical terms of the order that we introduced among versions of witnesses or length of objects by $<_p$ or $<_l$ may be obtained by requiring that real-valued functions P and L satisfy the following conditions:

$$\text{if } a =_p b, \text{ then } P(a) = P(b) \text{ and if } a =_l b, \text{ then } L(a) = L(b)$$
$$\text{if } a <_p b, \text{ then } P(a) < P(b) \text{ and if } a <_l b, \text{ then } L(a) < L(b)$$
$$P(a +_p b) = P(a) + P(b) \text{ and } L(a +_l b) = L(a) + L(b)$$

These conditions create a structural resemblance between numbers and their relationships on the one hand and entities of the real world and their relationships on the other.

Laws intervene at this very stage in pointing out and selecting relationships holding between properties of the phenomena and representing them *via* some numerical function. Considering the importance of the notion of (partial) isomorphism, we can conclude that it is not so much the nature of individuals that matters in such a structuralist reconstruction of the representational capacity of a scientific law, but the fact that some of the relevant relations between individuals are kept by the mathematical model.

In a word, laws are mathematical because we can know only the relational, dispositional structure of events and individual entities, and mathematics is the science of structures and forms. This thesis about the way mathematically formulated laws represent reality is also important to understand the nature of scientific change. As John Worrall has forcefully claimed (Worrall, 1989), what persists unchanged through scientific change is exactly such structures: Fresnel and Maxwell postulated an ether which has been "somehow" abandoned, but their equations are still part of the curriculum of every physics'student.

7. Conclusion

Here are two quotations which will offer me the opportunity for some conclusive comments on the questions that have occupied us till now: "... while structural relations are real in the sense that they are testable, the concept of unobservable entities that are involved in the structural relations always has some conventional element, and the reality of the entities is constituted by, or derived from, more and more relations in which they are involved." (Cao, 1997, p. 5)

In this interesting passage, the historian Cao specifies *the epistemological conditions* enabling us to arrive at a justified judgment about the reality of a theoretical entity ("more and more relations in which they are involved"). The main point of having laws is to specify such conditions in more and more precise way, but it is important to remark that by accepting Cao' thesis there is no need to deny the reality of the entities: relations (which are the object of laws) cannot be born without relata (the entities). In this sense, we should not confuse the mind-independent existence of the theoretical entities themselves with the network of relations, specified by theoretical and phenomenological laws, enabling us to discover them and to attribute them measurable properties.

The second quotation refers to a structural realist understanding of the way laws represent the world that is also an implicit answer to WQ: "To borrow from the ancient philosophical tradition, what I believe the history of science has shown is that on a certain very deep question, Aristotle was wrong and Plato – at least on one reading, the one I prefer – right: namely, our science comes closest to comprehending the real, not in its account of "substances" and their kinds, but in its accounts of the "Forms" which phenomena "imitate" (for "Forms" read "theoretical structure", for "imitate" "are represented by")" (Stein, 1989, p. 52)

Laws are mathematical because the relational structure of many natural system is partially isomorphic to the mathematical structure used to represent them, but the existence of such isomorphisms can only be explained by pointing to the "empirical origin"[12] of some fundamental mathematical concepts, from the geometrical forms that we extract from our perceptions to the experience of temporal succession, which is at the basis of arithmetical counting. The importance of understanding the origin of mathematical concepts with the help of the resources of cogni-

[12]By referring to "empirical origin" of the fundamental mathematical concepts I do not mean to exclude the possibility that such origin be, to use a turn of phrase dear to Spencer, "a posteriori for the specie but a priori for the individual".

tive science may open new paths to the philosophy of mathematics and of natural science in general (see Longo, 2002).

Acknowledgements. I would like to thank the audience at Copenhagen for the interesting questions raised during the discussion of the paper.

References

Barrow, J. (1988). *The World within the World.* Oxford University Press, Oxford.

Barrow, J. (1992). *Perché il mondo è matematico?* Laterza, Roma-Bari. (Why is the world mathematical?).

Bueno, O., French, S. and Ladyman, S. (2003) On representing the relationship between the mathematical and the empirical. *Philosophy of Science*, 69:452–473.

Butts, R. (1968). *William Whewell's Theory of Scientific Method.* Pittsburgh University Press, Pittsburgh.

Cao, T. (1997). *Conceptual Developments of 20th Century Field Theories.* Cambridge University Press, Cambridge.

Carnap, R. (1950). *Logical Foundations of Probability.* Chicago University Press, Chicago.

Cartwright, N. (1994). *Nature's Capacities and their Measurement*, chapter What Econometric can teach Quantum Physics: Causation and the Bell Inequality, pages 231–263. Oxford University Press, Oxford.

Deutsch, D. (1985). Quantum theory, the Church principle and the universal quantum computer. *Proceedings of the Royal Society*, A 400:97–115.

Hempel, C. G. (1958). The theoretician dilemma. In Feigl, H., Scriven, M., and Maxwell, G., editors, *Minnesota Studies in the Philosophy of Science*, volume 2, Minneapolis. University of Minnesota Press.

Kant, I. (1968). *Metaphysische Anfangsgründe der Naturwissenschaft*, volume IV of *Akademie Textausgabe.* Akademie Verlag, Berlin.

Longo, G. (2002). The constructed objectivity of mathematics and the cognitive subject. http://www.dmi.ens.fr/users/longo.

Pour-El, M. and Richards, I. (1989). *Computability in Analysis and Physics.* Springer Verlag, Berlin.

Stein, H. (1989). Yes, but... Some skeptical remarks on realism and antirealism. *Dialectica*, 43:46–65.

van Fraassen, B. C. (1982). The Carybdis of realism: Epistemological implications of Bell's inequality. *Synthese*, 52:25–38.

Wigner, E. (1967). The unreasonable effectiveness of mathematics in the natural sciences. In *Symmetries and Reflections*, pages 222–237, Indiana. Indiana University Press.

Worrall, J. (1989). Structural realism. *Dialectica*, 43:99–124.

HOW NATURE MAKES SENSE

Jan Faye

Traditionally the aim of science is partly associated with the discovery of things and properties, and partly identified with finding of laws of nature. Because laws are taken to explain why particular phenomena are what they are, or occur as they occur, science is interested in revealing as many laws as possible. Whenever we observe things and properties occur and behave in a regular way, we believe it is because laws of nature link things with other things, properties with other properties. A familiar opinion is that by linking things together laws of nature simply force things and events to subdue and obey. Laws tolerate no exception; laws are ontologically superior to things and events. Thus if it is impossible for anything to go against them, science can use laws to explain why sorts of things in connection with other sorts of things happen in the way they do. It is therefore quite common to consider the causal behaviour of individual things as a manifestation of some underlying causal laws. From an epistemological perspective we might get to know individual causes first, and then causal laws. But from an ontological perspective the opposite is the case: causal laws come first, individual causes next. In reality causal laws are more fundamental than particular instances of them.

Although this image of laws is prevalent, there are good reasons to resist it. First, laws might not be causal in nature; second, causes need not entail laws. The nature of laws and the nature of causality are logically distinct issues in spite of the fact that causal laws exist. As long as one regards cause and effect as succession of sorts of events, there is a tendency to see causation as a species of law-like connections, where laws are nothing but regularities in the behaviour of things. In contrast I have elsewhere argued in favour of a singularist theory of causation by rejecting the view that causes involve regularities of events (Faye, 1989, Chap. II and Faye, 1994). But by doing this I have not necessarily dissociated myself from a view of laws according to which laws of nature should be characterised as nothing but uniformities in

J. Faye, P. Needham, U. Scheffler and M. Urchs, (eds.), Nature's Principles, 77–102.

the pattern of things and events. Logically, we may still accept that laws are reducible to regularities.

The central issues are therefore whether our notion of laws contains something over and above the notion of regularities and, if it does, whether this is due to a mental abstraction based on our observation of regularities among events. In other words do laws of nature exist as mind-independent entities? And if they do, can the nature of laws of nature be different from the uniform manifestation of them?

To answer these ontological questions in the appropriate manner we must investigate whether or not all laws have real ontological counterparts in the sense that there always are nomic facts making law statements true or false. Is every statement that appears to be a law really descriptive of nature? First, I shall argue in line with other authors that we must distinguish between at least two different sorts of law statements. Second, I hold in opposition to them that one of these two sets of laws do not state any law of nature but must be considered as definitions, meaning postulates or language rules for formulating the other set of laws that expresses the real laws of nature. Third, I shall analyse what are the truth makers for statements about the real laws of nature, that is the ontological counterpart to real law statements.

1. Two main types of laws

As we want to explain the nature of laws, it might advance the matter to investigate the various possible types of laws to see whether they require different forms of explanation. One way of making a differentiation is the philosophical distinction between *empirical* laws and *theoretical* laws. The contrast is here between laws concerning observable entities and laws concerning unobservable or so-called theoretical entities. Then the empiricist approach would attempt to reduce theoretical laws to empirical laws; whereas the realist approach, accepting the distinction, would attempt to explicate empirical laws in terms of theoretical laws. But I believe that the empiricist distinction between entities that can and cannot be seen by the naked eye is epistemically uninteresting and irrelevant, and therefore that it cannot be restored to favour again as we are dealing with laws of nature (Faye, 2000). Whatever laws of nature are, they happen to involve visible as well as invisible entities.

Nancy Cartwright points to the distinction in physics between *phenomenological* and *fundamental* laws as different from the philosophical distinction. In physics the contrast is not between laws governing observable entities and fundamental entities. Rather, physicists make a distinction between laws that explain and laws that merely describe.

As she says, "In modern physics, and I think in other exact sciences as well, phenomenological laws are meant to describe, and they often succeed reasonably well. But fundamental equations are meant to explain, and paradoxically enough the cost of explanatory power is descriptive adequacy. Really powerful explanatory laws of the sort found in theoretical physics do not state the truth." (Cartwright, 1983, p. 3) A little later she continues, "I will argue that the falsehood of fundamental laws is a consequence of their great explanatory power." (Cartwright, 1983, p. 4) Unfortunately, Cartwright does not give a more precise characterization of the two types of laws. But she provides us with some further clues: "The causal story uses highly specific phenomenological laws which tell what happens in concrete situations. But the theoretical [*sic*] laws ... are thoroughly abstract formulae which describe no particular circumstances." (Cartwright, 1983, p. 11) Now it is difficult to see how fundamental law statements can, at the same time, be false and contain no descriptive content. For by assuming that fundamental laws are contingent, they must describe either some particular or universal facts, but then they are either true or false, or they don't describe facts at all, but then they are neither true nor false. Fundamental law statements do not describe particular facts, but neither can they describe universal facts, because if they did, they would sometimes be true. But, according to Cartwright, they are always false. Hence they cannot be true either. Nonetheless, she still thinks that the laws enable us, somehow, to tell the causal story. (Cartwright, 1983, pp. 55, 69, 161–162) How can false laws tell a true story? She realizes that there is a problem; she admits having no account of how fundamental laws manage to yield such a story. She does not know how such laws have explanatory force without being true.

Perhaps the puzzle should be taken as a sign that something has gone astray in the analysis she offers. Behind all her rhetoric the argument seems to be this: all quantitative laws of physics, like the law of gravity, have exceptions. (While claiming this, Cartwright does not distinguish between fundamental and phenomenological laws, but I take that what she has in mind is fundamental laws.) Therefore physical laws are in general false. The law of gravity is only true of a physical system that contains no other force than gravitation. Such a covering law is scarce because it holds only for very few systems. If physical laws are formulated by including their exceptions instead, then we will get *ceteris paribus* laws. But *ceteris paribus* laws are no true laws. For they hold only under special and idealized conditions. So without the *ceteris paribus* modifier physical laws are false, and with it they are not really laws. Therefore, *ceteris paribus* laws cannot figure in a covering law model of explanation in spite of the fact that they play a fundamental explanatory role in physics.

This is, I think, a fair description of Cartwright's view as it is presented in her book. But I also think that her argument contains a couple of flaws. Why are *ceteris paribus* laws not real laws? I cannot see why some of them shouldn't be counted laws. The conclusion that Cartwright draws about *ceteris paribus* laws fails to distinguish between being a fundamental law and being a causal law. It might be true with respect to fundamental laws that they are not real laws, but not true, I would say, with respect to causal or phenomenological laws. In my opinion, causal laws must in their complete formulation exclude exceptions by including nomologically relevant circumstances, which, because they leave out other potential causes, must at least be fixed for one particular event to be a cause of another. The nomologically relevant circumstances are those conditions that always have to be fulfilled so that one kind of event causes another kind of event. No event causes another event in every circumstance. A *ceteris paribus* generalization therefore need not, as we shall see, collapse to a Ramsey generalization. If it is a causal generalization, it does not. And if it isn't a Ramsey generalization, there is no generic distinction between causal laws with and without a *ceteris paribus* modifier. For what the modifier does is merely to restrict the domain of a causal law.

If *ceteris paribus* generalizations are not Ramsey generalizations, we still have a problem regarding how to diffentiate between a *ceteris paribus* law having only one instantiation, and a singular causal fact that does not entail a law. The suggestion that there might be cases of causality that do not involve a law is at risk, if the *ceteris paribus* clause can hedge in a domain so only one instance occurs in it. For example, Lee Harvey Oswald kills President Kennedy. Examples like this one have previously been offered as illustrations of causal connections that don't imply a law. Thus, if we open up for *ceteris paribus* laws, it seems quite obvious that one could formulate a law like: "Every person who has the same history and character as Lee Harvey Oswald's, and who has exactly the same relationship to his country and president as Oswald had to the US and Kennedy, will kill his President whenever he becomes in the same situation as Oswald." The consequence would be that every singular causal statement apparently supports a law. Either we say *ceteris paribus* generalizations are true causal laws, but then every case of causation would require the existence of a law; or we deny that every causal fact is an instance of a law, but then *ceteris paribus* generalizations cannot have the status of being genuine laws.

To avoid such a consequence I shall suggest one way out of the dilemma. The proposal is based on what can be regarded as properties belonging to kinds of entities and properties pertaining only to an individual. On the one hand, if a *ceteris paribus* generalization involves those properties that an entity possesses in virtue of being a certain kind, then we have a true law. If, on the other hand, a *ceteris paribus* generalization only relies on characteristics which an entity has because it is that particular individual, and not because it belongs to a class of entities, then we don't have a real law. With this distinction at hand, I see no reason to follow Cartwright's total rejection of *ceteris paribus* generalizations as true laws.

Now, I for my part think that most fundamental laws, in contrast to phenomenological laws, are neither contingently true nor contingently false, since they play a prescriptive role contrary to the descriptive one of causal and structural laws. It is their prescriptive character that makes fundamental laws well qualified to form the basis of the causal story. Apart from this essential disagreement I agree with much of what Cartwright says about fundamental laws and phenomenological laws.

I think that science operates with two types of laws that differ logically from each another. There are (i) *fundamental* laws, and (ii) *empirical* or *phenomenological* laws. The first kind is mostly prescriptive (some of them are, if you like, true by definitions), the second one descriptive. Both categories of laws contain various subcategories: Among the fundamental laws we have *theoretical* laws that express relations between quantities, but among them we also find conservation laws, symmetry principles and theoretical postulates. Empirical or phenomenological laws include, for instance, *causal* laws, *structural* laws and *functional* laws. In the rest of this paper I shall mainly focus on theoretical laws in contrast to causal laws. The pronounced difference is that theoretical laws don't tell us how physical objects in their domain actually behave, they state how two or more quantities are interrelated, whereas causal laws provide us with an account of what makes physical objects behave as they do. As we shall later argue causal laws can be reached through a certain sort of generalization, not from regular conjunctions or successions of the same kind of events as the regularity account would have us to believe, but from singular causal facts. Theoretical laws, on the other hand, are the result of definitions or rule-giving prescriptions for how causal laws should be formulated in a model in order to explain.

2. Theoretical Laws

In contrast to Cartwright I shall argue that theoretical laws don't express facts, and therefore that they don't contain *ceteris paribus* clauses, whereas causal laws state facts, and by doing so they always contain a *ceteris paribus* operator. Thus, law statements without exceptions are the theoretical laws; they don't describe any law of nature, whereas law statements with exceptions deal with the real laws of nature. But, then, if we deprive theoretical laws of their facticity what status do they have?

Besides Cartwright several other authors have entertained the idea that science contains different kinds of laws without being able to explicate the demarcation. Kuhn's paradigms, for instance, comprise something he calls symbolic generalizations. These may have symbolic form like Newton's second law of motion, $F = ma$, or Ohm's law, $I = V/R$; or they can be expressed in ordinary words: "elements combine in constant proportion by weight," or "action equals reaction". Kuhn contrasts symbolic generalizations with what he calls exemplars that illustrate the symbolic generalizations. According to Newton's second law such exemplars would be the case of the free fall described by the equation $mg = md^2x/dt^2$; the inclined plan by $md^2x/dt^2 = mg\sin\theta - mg\cos\theta$; the simple pendulum by $mg\sin\theta = -mld^2\theta/dt^2$; a pair of interacting harmonic oscillators described by two equations, where the first can be written $m_1d^2x_1/dt^2 + k_1x_1 = k_2(x_2 - x_1 + d)$; etc. These exemplars correspond to various models by which Newton's second law can be used to derive phenomenological laws. Symbolic generalizations, however, do not say anything about concrete situations. According to Kuhn, "they function in part as laws but also in part as definitions of some of the symbols they deploy. Furthermore, the balance between their inseparable legislative and definitional force shifts over time." (Kuhn, 1962, p. 183) But how can symbolic generalizations be both descriptive and prescriptive? Kuhn admits that this requires a further analysis because our commitments to a law of nature are very different from our commitments to a definition. Laws are corrected little by little, whereas definitions, being tautologies, are not corrigible at all. I think that Kuhn has isolated the symptoms, but he has not served us with a diagnosis.

His claim that laws are acting like definitions flies in the face of the realist opinion. But the claim is not new. Already Henri Poincaré argued about establishing the equality of two forces that in such a case Newton's second law of motion "ceases to be regarded as an experimental law, it is now only a definition." (Poincaré, 1952, p. 100) And he was not alone. In his excellent treatment of the origin and nature of Newton's laws of motion Brian Ellis writes: "In the tradition that has succeeded Mach,

Newton's second law of motion has been widely regarded as a definition of force." (Ellis, 1965, p. 52) In addition, based on historical studies too, Norwood R. Hanson held that Newton's laws of motion have had many distinct uses. Considering the practices of physics, he said, one will discover that 'the law of inertia', 'the second law of motion', and 'the law of gravitation' all stand for 'umbrella-titles'. By covering a family of roles their mathematical counterpart can be taken to play definitions, a priori statements, heuristic principles, empirical hypotheses, rules of inference, etc. (Hanson, 1958, p. 112) Hanson's suggestion builds on Wittgenstein's idea of a language game. It depends on the historical, theoretical or experimental context which of these possible roles a law statement actually appears to have. Neither of these distinct uses can be said to be the correct one. Also Stephen Toulmin thinks of laws of nature as definitions holding that "laws of nature resemble other kinds of laws, rules and regulations. These are not themselves true or false, though statements about their range of application can be." (Toulmin, 1953, p. 89)

In the following I shall explain the source of the various uses in a different way while I also raise doubts about several of them. First, I shall argue that theoretical laws are definitions; they don't have a descriptive content. Secondly, I shall argue that theoretical laws don't have a priori status, in spite of the fact that they are definitions. In his discussion of Newton's laws, Hanson is a typical representative of a person who conflates analyticity with a priority. Thirdly, I shall argue that theoretical laws are neither contingently true nor contingently false.

Theoretical laws don't serve as descriptive definitions but only as prescriptive ones. A descriptive definition states the actual meaning of the definiendum by claiming that the definiens is synonymous with the definiendum. It is a claim like 'All bachelors are unmarried men'. Therefore such a definition is true or false as a description of the actual use of the term 'bachelor', and its truth-value can be known a priori as long as one knows the meaning of that term. A prescriptive definition, on the other hand, stipulates that henceforward a well-known term should be used in the stated sense of the definiens, or such a definition introduces a completely new expression. This type of rule making definition will be more or less adequate but not true or false.

The last statement immediately raises a fundamental question. If theoretical laws are neither true nor false, how is it possible to account for their use in arguments that are supposed to be truth-preserving? I take it that different notions of truth are at stake here. Truth is sometimes more than nothing but the truth. I believe we ought to distinguish between *analytic* truth and *real* truth. An argument, or an inference, is valid if and only if no argument with the same logical form has true premisses and a false conclusion. This definition of validity is true in every possible world. This kind of truth must in general be distinguished from the one we attribute to the premisses and the conclusion. It is an abstract and formal notion of truth. A sentence like 'A bachelor is an unmarried man' is true in the same analytic sense. But that sentence, as we just argued, is also true in another, real sense: it is contingently true with respect to our present use of the term 'bachelor'. So by saying that prescriptive definitions are neither true nor false I mean that they are neither contingently true nor contingently false. What a person does at most when he proposes such a definition is that he recommends to the linguistic community that what is said by the very expression should from now on be considered to be analytically true, i.e. true in all possible worlds, as a rule for description in the actual world. It is only in this sense that their use in arguments may be truth-preserving.

What distinguishes laws of nature from language rules is that laws are concerned with the natural world, whereas rules specify a certain linguistic or social behaviour. So why should, say, Newton's *laws of motion* be rule making definitions and not literally descriptive sentences stating empirical facts or contingent states of things? I shall provide four arguments:

First, if fundamental laws like Newton's laws of motion were empirical generalizations, or could be furnished with an empirical content in some other ways, Cartwright would be right in claiming that the laws are false. This is a point on which Ronald Giere agrees. Therefore he holds the view that Newton's laws are definitions. Such laws, he says, are to be interpreted as providing definitions of various models (Giere, 1988, pp. 77–78, 84). I don't entirely agree with Giere's handling of models, but this is inessential (Faye, 2002, Ch. 8).

Second, I think Brian Ellis is correct when he argues that forces are queer entities because they "are not like other theoretical entities, such as atoms, or genes, since the existence of atoms or genes is not entailed by the existence of the effects they are supposed to produce." (Ellis, 1963, p. 188) Instead "the action of forces", he says, is "supposed to explain certain patterns of behaviour, the occurrence of these patterns is considered a sufficient condition for the existence of the precise force required to produce them." (Ellis, 1965, p. 31) In other words the existence of forces are entailed and entailed by those effects they are supposed to produce, which means that forces are very different from ordinary causes in the sense that ordinary causes, but not forces, are logically distinct from their effects.

Third, Newton's three laws of motion are not the only appropriate formulations of classical mechanics. There are alternatives, which do not introduced mechanical forces, for example, Lagrange's and Hamilton's formulations. According to Newton's second law of motion, $F = dp/dt$, the movement of a particle is explained by a force function F. Joseph Louis Lagrange, however, proved that the same behaviour could be explained by a new function L, the so-called Lagrangian, defined in terms of kinetic and potential energy, $L = p^2/2m - V$. The new law becomes $d/dt(\delta L/\delta \dot{q}_j) - \delta L/\delta q_j = 0$, where q represents the generalized coordinates. A further reformulation of the classical equation of motion is due to William Rowan Hamilton. He introduces another function H, the so-called Hamiltonian, where $H(p, q, t) = \Sigma \dot{q}_i p_i - L(\dot{q}, q, t)$; and a set of $2n$ first order equations of motion: $\dot{q}_i = \delta H/\delta p_i, -\dot{p}_i = \delta H/\delta q_i$. Thus the movement of a classical system can be explained either in terms of the Newtonian function F, the Lagrangian function L, or the Hamiltonian function H. What can we say about these alternative formulations?

The problem is twofold: On the one hand, if any of the above formulations of the laws of motion had had a true descriptive content, we would have expected that the particular function in question would have made a difference in the observational consequences (unless, of course, some of the content is ontologically superfluous.) But all three formulations are empirically equivalent. On the other hand, instead of denying them any descriptive content, one often sees attempts to argue that they are so equivalent because they share the same descriptive content.

The basis for these two suggestions should be elaborated a bit further. If the various formulations are equivalent, this shows, one may say, that the descriptive content is in fact the same, not that there is no such content. Or, if instead of equivalence we only have implications one way, then descriptive content is reduced in the direction of implication; i.e. what is implied retains some of the descriptive content without introducing new content. But what kind of equivalence yields the same descriptive content? In case we attribute a certain descriptive content to two or more sentences, this content must be either the same or different. This requires that we have some semantic criteria to judge the commonality or the contrariety of their content. But what these criteria are is rather obscure.

If one argues, as the positivists did, that the theoretical equivalence of two theories reduces to their observational equivalence (because the theoretical terms get their meaning from the observational terms), then the descriptive content is the same for all three formulations. One could also maintain a holist-theory of meaning, claiming that the meaning of the theoretical terms is fixed entirely by the role they play in their respective theories. Two theories are theoretically equivalent just in case there is an appropriate structural correspondence between the two theories such that one theory can be obtained from the other in virtue of a simple interchange of term by term. Again one arrives at the conclusion that the descriptive content of all three formulations is the same. However, already some time ago Lawrence Sklar pointed out that these two candidates for the semantic equivalence of theoretical expressions face a series of problems. So the idea of what counts as the same descriptive content is not in any way unproblematic (Sklar, 1982). A third possible candidate for a criterion would be to say that two expressions are semantically equivalent if and only if they are synonymous, that is if and only if they are intensionally equivalent. If this is one's choice, then the above formulations would not be semantically equivalent (the functions have different dimensions), and therefore they would not have the same descriptive content. But this is not satisfactory either. The up-shot of this brief discussion seems to be that there are no convincing criteria for semantic equivalence of theoretical terms and, depending on one's choice, it turns out that Newton's, Langange's and Hamilton's formulations have or have not the same descriptive content. We have no objective grounds to say which of them, if any, is the correct one in the sense of picking up a real entity. In my opinion this is not because they share the same descriptive content, but because of a bad habit of calling theoretical laws for laws of nature. Rather they should be named definitions or language rules.

Finally, fourth, the mathematical structure of theoretical laws often allows us to talk about phenomena that may or may not exist. Take negative energies, advanced potentials, Higg particles, magnetic monopoles, etc. All second order differential equations have negative solutions that are discarded as useless since we do not consider them to give us a literal description of anything. This feature can easily be explained if we rather think of theoretical laws as prescriptive rules of descriptions. Like natural language rules, theoretical laws provide us with a wealth of descriptive possibilities that may never be used to state or describe concrete facts.

We should note that there is no problem for the suggested account of theoretical laws as definitions or language rules, if someone were to argue that such laws imply counterfactuals. All expressions that can be defined according to rules will entail counterfactuals. For instance, Sunday is defined as the day after Saturday. This definition admits the counterfactual statement: 'If it had not been Saturday today, it would not be Sunday tomorrow.' A definition stipulates a necessary connection between definiens and definiendum; and since this form of analytic necessity is stronger than a counterfactual necessity, the former entails the latter, but not vice versa.

Furthermore, theoretical laws like Newton's three laws of motion cannot be a priori true statements. For had they been a priori true, it would be inexplicable why the Greeks thought that uniform motion was caused by an external force, and the Medievals thought that rectilinear motion was due to an internal force. Also, it would make any later replacement of Newton's theory with that of Einstein's absurd. At a time there were at least as good evidence for the Aristotelian theory of motion as there is for the Einsteinian theory today. But the fact that Aristotle's theory of motion was first turned down in favour of Newton's theory, and then Newton's theory was replaced by Einstein's theory does not tell us much about their logical status.

What history proves is that earlier definitions of motion were not adequate as a proper account of all kinds of motions. In his laws of motion Aristotle distinguished between natural and unnatural motion, and he then set up his force law. Natural motion needs no explanation; what needs one is unnatural motion and the change of natural motion: "All movement is either natural or enforced, and force accelerates natural motion (e.g. that of stone downwards), and is the sole cause of unnatural." (Aristotle, *De Caelo*, 301b 20–24) Of course, the Aristotelian force law stemmed from the sensations accompanying muscular exertion when we pull, push or lift to get an object to move, and the observation that if one stopped tugging and pushing the object it would not keep on moving. Nevertheless, Aristotle himself considered his law to be a definition.

In the Middle Ages another consideration made by Buridan and his followers started to change the situation into one where this definition was no longer taken to be adequate. In the process of creation, God had implanted an internal force, an impetus, into the planets that allowed them to move after his direct interaction had ceased. God didn't have to keep himself busy by pushing the planet around all the time. By analogy the same is the case when we throw a stone, or shoot an arrow. But one remarkable difference remained. The planets circle uniformly and continuously around the Earth, but the stone or the arrow comes to a halt very quickly. Again experience and conceptual reflections played together. Here on Earth things come to rest quickly because they meet some resistance. The less the resistance, the longer it takes things to stop. A stone being thrown on sand is observed not to move as long as one on ice. One can therefore imagine that the stone would move on continuously without any resistance like the planets. The law of inertia is about to be born. Also pushing a stone on sand requires more force than one on ice. So one needs force to overcome resistance. Newton's third law of motion is about to be born. And, finally, a stone at rest requires some kick, push or pull for it to move, the stronger a force the faster it moves. Newton's second law is about to be born.

Three times I said "about to be born." For no one has ever experienced a body moving in a straight line with a uniform velocity. The planets orbit in circles, or in ellipses, and not uniformly over the year, and earth-bound things move in parabolas. What was still missing to complete the birth was a conceptual move away from Buridan's idea that the external force is transformed into an internal force. Nothing empirical could produce such a change. An external force can be measured in virtue of its effects, whereas an internal force would not have any observable effect. Descartes was one of the firsts, if not the first, to state the law of inertia. Everything remains in its state in which it is unless some external agency causes it to change that state. This also applies to motion where a body must retain its speed and its direction of motion unless some external agency causes a change of them. Descartes came to this conclusion from his idea that God, as an immutable being, had created the world so that the total motion in the universe is conserved, or rather the product of the velocity and the quantity of matter is constant. The impetus had become momentum. In making the world God put everything into motion once and for all, and by doing so he had given everything a certain impetus, or momentum, which could change in a body only if it was accompanied by an equal and opposite change in another body.

From here Newton took over. What was needed was a new force law saying what causes the impetus or momentum to change, a force law that could replace the Aristotelian law. In Newton's own wording the new law became: "The change of motion [motus: *i.e.* momentum] is proportional to the motive force impressed; and is made in the direction of the right line in which that force is impressed." (Newton, 1946) What is important here is the word 'motive force.' Brian Ellis has convincingly argued that it has a meaning close to the primitive concept of a push or a kick rather than the strength of a kick at any given instant. (Ellis, 1965, p. 36) So the motive force is the total force an object receives during an impact with another object.

Thus, Newton's laws are not a priori true but they still function as explicit definitions of natural, force free, motion and unnatural, force impressed, motion. But how can a statement that is analytic but not a priori be known to be true? For the criterion of an analytic formula is that its negation entails a contradiction, and contradictions can be grasped a priori. A statement need not, as argued earlier, be synthetic for its validity to be recognized a posteriori. We can reject a synthetic sentence as false because it contradicts other synthetic sentences; and we can decline to accept an analytic expression as valid by replacing it with another analytic expression, because the latter is considered to be more adequate than the former. I hold that Newton's laws of motion, like other theoretical laws, are analytic but known only a posteriori to be valid.

As rule making definitions theoretical laws are neither contingently true nor contingently false. Instead they are valid within a certain domain, because they are used to set up a descriptive vocabulary for that domain. But their analyticity is not recognized without experience being consulted. While theoretical laws provide us with prescriptions for how we must causally describe concrete physical situations, we experience them as valid or adequate. If they fail to act as rules of description, we see them as invalid or inadequate. It is through their power of guiding our description of nature that we eventually come aware of the analytic status of theoretical laws. Thereafter we consider them as valid as long as it takes. Thus if the negation of a theoretical law no longer leads to a contradiction, it is experience only which can reveal it for us. First when evidence is offered to support the negation, it would be dismissed promptly. But if the evidence cannot be explained away, if the evidence shows itself recalcitrant, the scientific community can do one thing or the other. Either the community can restrict the domain in which the law is valid, and introduce a new set of rules for describing things outside this domain; or it can replace the law entirely with a new set of rules.

The prescriptive character of Newton's laws of motion explains why so much science and technology are not developed without applying these laws. For if they are neither contingently true nor contingently false, nor approximately so, they might still be found adequate as rule-making definitions within a certain domain. When we are dealing with slow moving, macroscopic objects, far from strong gravitational fields, these laws are as useful as ever. Still being valid in a domain also constitutes the reason that school children have to learn classical mechanics today. Indeed, it might be argued that since a causal law often can be expressed in terms of theoretical laws, such a law cannot be maintained to be true unless the theoretical laws post certain ontological commitments about the nature of the causal nexus. If that is correct, then theoretical laws must be true or false too. The argument fails, I think, partly because theoretical laws are not causal statements in themselves, and partly because a causal story can be told by means of different theoretical laws.

A theory like Newton's mechanics contains a vocabulary and a set of rules, that is his three laws of motion, which lay down how the vocabulary is defined. So his laws provide the physicists with the linguistic apparatus to describe models of concrete cases and thereby allow them to come up with a causal story or perhaps even formulate a causal law. But a scientific theory also contains a number of theoretical laws, which don't apply to the phenomena. Fitting phenomena to theory is therefore not a simple process. What complicates it is the abstract and idealized character of theoretical laws. Often there are many steps to take from observations and experiments to theoretical laws and back again to the world. There is no direct route from theoretical laws to the phenomena, but it always goes via physical models. To move from phenomena to theory requires entry rules, and returning from theory to the phenomena needs exit rules. The entry rules don't have their own canonical name, but the exit rules do: they are called explanations and predictions. We may, however, call entry rules for interpretations. But how theory and phenomena hang together in details is quite a different story, which will not be told here.

3. The regularity account of laws

What then is a law of nature? Is a law of nature identical to a regular and recurrent pattern of the same kinds of events? Indeed, if something should count as a natural law, it seems at least to be a requirement that we are able to observe similar events happening regularly, perhaps under certain nomologically relevant circumstances. It is the law that makes the regularity possible. A law of nature simply entails regularity. But does the entailment also hold the other way around? Many convincing arguments have been produced against the notion of laws as a mere association or a succession of kinds of events. As a first account of laws as regularities a statement p is usually said to express a law of nature if, and only if, the following conditions are satisfied: (i) p is universal in its scope; (ii) p is true everywhere and everywhen; (iii) p is contingent; and (iv) p contains only non-local empirical predicates and logical connectives and quantifiers. These constraints leave us with a formulation of p in terms of a material conditional: $(x)(F(x) \supset G(x))$, where F and G are empirical predicates. Such a statement meets all four conditions. The idea is to get rid of all accidental generalizations that are restricted in scope, say, all spheres of gold on earth are less than one kilometre in diameter, or all Conservative prime ministers of England between 1903 and 1928 have names beginning with B. Nevertheless, several things indicate that the above extensional formulation excludes some, which we consider as laws, but also includes uniformities, which we don't regard as laws.

Let us consider the first thing first. The suggested formulation of p has to be modified if such an expression should not exclude probabilistic associations. For many laws can only be expressed in terms of probabilities; that is, an F has a certain probability of being a G. The most obvious modification therefore becomes $(x)(F(x) \supset P(G(x)))$. The formulation seems to express some law-like connection very well: for all x, if x is a herbicide, then x will kill a certain percentage of the weed when it is sprayed with x. The modification creates, however, a major problem. It includes instances, which would not be regarded as a law. For instance: for all x: if x has gravity, then there is a certain probability that x is a human being. Undoubtedly this conditional is true but it is not a law. Rather we would say that it is accidental that some bodies with gravity are also human beings. Apparently the above expression fails to state a law whenever no causal fact is involved.

Now, other arguments seem to show that the regularity account misses some genuine laws of nature. Some laws may be very restricted in space and time; therefore, the quantification cannot range freely over all x in every time and every space. But also here the adherent of the regularity account can modify his formulations so that it ranges over some intervals in time or space. Strange counter-arguments, like Michael Tooley's example with Smith's garden, are no better or no worse than for other accounts (Tooley, 1997, p. 686). For if all kinds of fruit trees in Smith's garden carry only apples, and all kinds of fruits taken into the garden turn into something other than fruits, it is either because some spell is cast on that particular garden, or because it is a law with only one instance. In the first case, where it isn't a law, there is nothing for the regularity account to explain. In the second case, where it is a law, the law-likeness may exist for two reasons: some law operates only at this very limited space-time point, or some law consists of the fact that Smith is the owner of the garden. Every account has problems with the first form of law, unless it introduces some super law explaining why, at this particular space-time point, fruits act as they do. The second form of law exists in virtue of a relation between Smith and the fruits in his garden. What would make the case law-like would be that one of the relata consists of a combination of general properties that, by accident, Smith was the only person in the world to possess. But such a combination of general properties raises no particular difficulties for the regularity account. Of course, it becomes a problem for the regularity account, as for any other account, whenever we have to justify a belief in such a law.

A serious problem is what the extensional formulation includes. Already Frank Ramsey pointed out, using the example about the Conservative prime ministers, that what seems to form an accidental generalisation can always be put into terms, which are not limited to any particular portion of space-time (Ramsey, 1978, p. 130). One could simply say: 'All Conservative prime ministers of a country with 40,000,000–50,000,000 inhabitants, whose capital is called 'London' and has 7,000,000 inhabitants ... at a time when that country has between 2–27 years previously lost a queen who has ruled for 64 years ... have their names beginning with B.' The statement meets all four conditions for being a law. Yet, we have every aversion to elevating such uniformity to a law. If we include sufficient details into a generalisation and describe them in non-local terms, we can generate infinitely many universal statements from an accidental generalisation. How do we distinguish between these Ramsey generalisations and real laws of nature?

In support of the general feeling that the four conditions fail to pick out only that group of generalizations that really express laws of nature, one usually refers to an argument first put forward by William Kneale. Suppose a whole race of ravens has white plumage because it had been living in permanently snowy regions for a long time. Darwin's theory of evolution provided us with good reasons to think that something like this could happen. Hence, if the regularist were right in his view that the actual uniformity constitutes a law, he would have to oppose the physical possibility of white-feather ravens.

Popper has produced an identical argument (Popper, 1959, Appendix x). Suppose under ideal conditions a moa, an extinct bird from New Zealand, could live sixty years or longer according to its biological structure. Furthermore, suppose no moa has ever become fifty years old or more, because before reaching this age a deadly virus in the environment of moas dispatched each and every one. Consequently, a statement that 'All moas die before they become fifty' would be true, although it does not express a law of nature. For it is due to accidental or contingent conditions that moas didn't live longer.

The objection fails, I think, to be of any menace to the regularity theory. It springs from a too narrow view of laws of nature. How can we be so sure that general laws exist pure and simple? Even if laws exist as something different from their manifestation, they may still be able to interact favourably or to interfere unfavourably with each other. By acting together quite new effects may be created, effects that cannot be associated with one general law, or effects of one law may be cancelled by the influence of another law. Moreover, we are inclined to think of laws of nature as independent entities that can be combined or recombined like Lego bricks. Simple laws can combine into more complex laws. But just because the explanation of a phenomenon involves a combination of several independent law statements, we have no reason to believe that corresponding to each such phenomenon there is a single law of nature. Nor do we have good reasons to believe that the phenomenon arises as a result of an interaction between these laws that always remain the same. Believing that this is the case rests on what I will call the picture theory of laws. The view seems to require that there be a higher order law of nature explaining how the simple laws of nature interact in combination. After all the view does not escape the introduction of a hierarchy of complex higher order laws, perhaps of infinite many orders.

The above objection, however, takes also for granted that it is always possible to distinguish between accidental, or contingent, conditions on the one hand and nomical conditions on the other hand. According to the regularist, who doesn't think that laws have content different from their uniform manifestation, it is impossible to separate *universally* contingent conditions and nomical conditions. Thus, the regularist could claim that statements 'All ravens are black' and 'All moas die before they become fifty' do express laws of nature, since both sentences are abbreviated versions of more specifiable uniformities. One could hold the following statements as true: 'All ravens not living in permanently snowy areas are black' and 'All moas infected with a certain virus die before they become fifty'. Instead of representing, say, the former law by a simple formula like $(x)(R(x) \supset B(x))$, one should therefore express it as $(x)(y)(R(x)\&{\sim}S(y) \supset B(x))$, where the quantifier y runs over all circumstances, which do not have the property S of being permanently snowy. Such a formulation seems quite acceptable as an expression of a law.

The opponent of the regularity account seems to have a real problem here, unless he grants that the interplay of different laws might give rise to new emergent laws. For the regularist can always argue that we have a law as long as some uniformity is available. He does not have to assume that this regularity can be reduced to the interaction of two, more basic uniformities, if neither is ever manifest. In other words, he can deny that the uniformity appears as a result of the addition of two separate uniformities. The opponent operates with an unproven assumption that the world contains only a few basic laws that in various combinations provide all actual uniformities.

At first sight the regularist's respond does not address the essential point behind Kneale's and Popper's argument. For they want to advocate the existence of unrealised physical possibilities. But if it is a law of nature that moas die before fifty, then it is physically impossible that moas should live longer than fifty. The regularist must deny that it is physically possible for a moa to become sixty (Armstrong, 1983, p. 18). In my opinion, however, nothing forces the regularist to draw such an unwelcome conclusion. Something may be physically possible either because it would be in accordance with some actual law, or because it could be realized by a possible law. In other word, the regularist can make a distinction between actual and possible laws, saying that L^* is a possible law, if L^* contradicts an actual law L but no other actual laws. This means that the uniformity given by $(x)(y)(R(x)\&{\sim}S(y) \supset B(x))$ above is an L in the actual world if, and only if, every case in which an instance x of R satisfies B happens to be the circumstances non-S and never S. A possible law L^* would therefore be a uniformity given by $(x)(y)(R(x)\&S(y) \supset {\sim}B(x))$, where the circumstances mentioned in the antecedent are in conflict with the actual circumstances, and hence what the consequent says is in conflict with what is actually realized. Nevertheless, such a possible law does not contradict other actual laws like those expressed by Darwin's evolutionary theory.

Against the regularity account of laws Popper has raised another dubious argument (Popper, 1959, Appendix x). According to him, the regularity account sees the primacy of repetitions as a kind of justification for the acceptance of universal laws or, at least, as what causes a subjective expectation of the validity of the laws. Nevertheless, such view is untenable because all repetitions of events we experience are based on a similarity of features among different events, and similarity is relative: every event is similar to every other event in some sense, and dissimilar to all others in some other sense. But this means, says Popper, that subjective interests and psychological points of view are logically prior to repetition. The substance of his objection is, however, the same as that of Wittgenstein's against ostensible definitions. So the argument fails for the same reason. For the repetitions we experience are not grounded on arbitrary chosen similarities. They are based on such features that make it possible for us to refer correctly to particular events by using the same term. It is our ability as competent speakers to put names on things and events that causes us to see repetitions whenever we observe a recurrent pattern of uniformity between particular things and events to which the same names apply. But were we not able to extract visual resemblances from particular things before possessing a concept of that sort of thing, the invariant usage of common nouns and predicates, being necessary for all communication, would not be possible. The application of terms would instead take place in a complete random order.

Popper might not deny this: however, he believes at least that such a fact would not help us to use regularities in justifying statements about laws of nature. For, as an empiricist he claims, not only do universal laws transcend experience but also so does the content of singular statements. The latter goes beyond experience "because the universal terms which normally occur in them entail dispositions to behave in a law-like manner" (Popper, 1959, p. 425), whereas the former goes beyond experience partly because they transcend any finite number of their observable instances, and partly because they contain universal terms in their formulation. But such a claim is utterly mistaken. First, by saying that universal terms entail dispositions to behave in a *law-like* manner, Popper presupposes what he has to prove: he has yet to show that dispositions are different from mere regularities and repetitions. Thus, universal terms might just entail dispositions to behave in a regular and repetitious manner. Second, the assumption that all universal terms are in principle disposition terms, as Popper argues, is completely unjustified, as well as the assumption that our use of such terms is determined by the acknowledgment of dispositional features. We apply universal nouns and predicates like 'mountain', 'blackbird', 'grass', 'blue', 'black', and 'green' quite independently of any knowledge of dispositions. Even if we grant Popper that things which are blue, or green, have a disposition to reflect blue, or green, light, knowing such a disposition would not have any effect on our capacity to use these predicates correctly. Most people master colour predicates without having the slightest knowledge about light and optics. In even greater contrast to Popper, I also hold that singular statements, containing general terms for so-called theoretical entities, will not transcend experience since such entities like atoms, electrons, etc., are observable though not perceivable.

The regularity account stands up against the above objections. But this does not prove its validity. For it is not valid. Where his opponents count too few uniformities as genuine laws, the regularist himself counts too many. The latter is unable to draw a fundamental distinction where one should be made; that is between Ramsey generalizations and truly law-like generalizations. The regularity account of laws fails for the same reason as the similar account of causation. Ramsey generalizations do not entail counterfactuals, only law-like generalizations do. The following statement would not be true: 'If Mr. Jones had been the Conservative prime minister in a country with 40,000,000–50,000,000 inhabitants, whose capital is called 'London' and has 7,000,000 inhabitants ... at a time when that country has between 2–27 years previously lost a queen who has ruled for 64 years ..., then his name would have begun with a B.' In contrast, counterfactual statements like 'If this ivory gull had been a raven living in areas not permanently covered by snow, it would have been black' and 'If this emu had been a moa, it would never become more than fifty years old' can be counted as true. Consequently, what marks genuine laws is their support of counterfactuals.

4. Causal Laws

Empirical laws appear in various forms. We have, among others, laws of coexistence and laws of succession. I shall nevertheless hold that laws of coexistence supervene often, although not always, on laws of succession. Take Boyle-Mariottes law which says that $PV = nRT$. This macroscopic law states the coexistence of pressure, volume and temperature of a gas. We know, however, that both the pressure and the temperature are macroscopic properties that result from the causal behaviour of the microscopic molecules of the gas. This example does not count as an argument but it shows that macroscopic laws of coexistence may sometimes supervene on microscopic laws of succession. But even if not all laws of coexistence supervene on laws of succession (and I don't think so) they have still to support counterfactuals.

Causal laws are *ceteris paribus* laws. This claim will be argued in this section. In case we observe the same kinds of events causing each other under the same kind of circumstances we have a causal law. Causal laws can be expressed by using the verb 'cause' or any related word that the verb can replace. The following statements are examples of causal laws: 'Cyanide kills people'; 'Smoking causes cancer'; 'Aspirin relieves headache'; 'The change of seasons is due to the earth's orbit around the sun'; 'Raising the temperature of a gas with constant volume increases its pressure'; 'The bombardment of uranium with neutrons splits the atoms and releases energy'.

In general we can write causal laws as 'C causes E', where C and E designate different sorts of events, properties, etc. Some of these laws are not deterministic in the sense that, say, it is a law that smoking causes cancer, but not every case of smoking necessitates cancer. Individual causes are, I hold, deterministic, but causal laws need not be. It seems to indicate that causal laws cannot be understood as a simple quantification over singular causal facts. Above I marked off causes that support a law from causes that don't. The argument was that a *ceteris paribus* generalization is possible as a law if the causal nexus between the events in question exists because of some generic properties of the entities involved; whereas it is not possible if the causal nexus holds between individual properties. So in case we have a causal law, we treat the causal entities as kinds and not as individuals. This raises, indeed, the question about the existence of universals and whether the causal law is a necessary relation between such universals.

The first problem we have to face is whether causal laws are merely generalizations of singular causal facts. In case they are, we seem to have

CL 'C causes E' is a law of nature if, and only if, c_1 causes e_1, c_2 causes e_2, ..., and so on.

The delicate issue is that if we are unable to understand a causal law as a generalization of singular causal facts, we can never get to the causal law by induction from observation of alleged cases of the law. Causal laws imply singular causal facts, but never vice versa. So where the regularist believes that causal laws are nothing but regular successions of similar *non-causal* facts, I believe that causal laws are regular successions of similar *causal* facts.

Is it possible to get from singular causes to causal laws if general probabilistic laws are at stake? I think it is. It is a law that smoking causes lung cancer, but certainly not every case of smoking gives rise to that disease and other events can bring about it too. First we are interested in what is the basis in reality for this law, what makes a statement of such a law true; and next we wish to know how it can be warranted. The ontological ground for being a law that smoking causes lung cancer is that some persons get lung cancer because they smoke, in spite of the fact that most people don't get lung cancer from smoking. Obviously this variation is due to the fact that either some relevant circumstances must be realized before individual cases of smoking causes lung cancer, or particular cases of smoking do not invariably determine lung cancer under the same kind of circumstances. I have already stated my misgivings against the latter alternative. Singular causes are not indeterministic by themselves; they always determine their effect. I shall therefore take the first alternative. The upshot is that any generalization of singular causes must also be quantification over the relevant circumstances. Thus, we have that

CL* '*ceteris paribus*, C causes E' is a law of nature if, and only if, $(x)(y)(C(x) \& K_J(y)$ causes E(x)), where y runs over all nomologically relevant circumstances K_J.

From an ontological perspective it is possible to specify a causal law of nature as a generalization of singular causal facts, but from an epistemic perspective the possibility of justifying such a specification is not always open.

In those cases we can isolate all the nomologically relevant circumstances through extensive experimentation, the way is right open for a justification of the causal law as quantification over individual causes. The problem is that the experimental procedure is not available in many cases. Instead of moving bottom-up, the scientist must now move top-down. This makes the warrant much more difficult and sometimes even practically impossible. A scientist starts out with an observation that cases of lung cancer have increased over a period of time, and his natural suspicion is that this is caused by something that people get into their lungs. Smoking is an open and shut candidate. A lot of lung cancer patients are then partitioned into smokers and non-smokers, and the statistical calculation show that ten times more smokers get lung cancer than non-smokers. Thus we take the probabilities, Prob $(L|S) >$ Prob $(L|{\sim}S)$, as evidence that it is a law that smoking causes lung cancer. But these probabilities are not evidence for the claim that this law can be considered as quantification over singular causal facts. Neither do they necessary reveal that smoking causes cancer is a probabilistic law and therefore that causes can be reduced to probabilities. They would if it was impossible to raise the probabilities to one. In case further investigations can raise the probabilities by including other nomologically relevant factors, it will not be a probabilistic law. As long as the probabilities are mixed, *i.e.* their reference class is inhomogeneous, we have not yet been able to specify every nomologically relevant circumstance; thus, we cannot justify **CL***. We stay without any warrant that this particular law is not one which goes beyond the collection of singular causal facts. But whenever we are able to turn the mixed probabilities, Prob $(L|S)$&Prob $(L|{\sim}S)$, into pure ones, Prob $(L|S.K_J)$&Prob $(L|{\sim}S.K_J)$, we have every reason to believe that first become one and the second zero. And if we can supply a general argument, we have by that not only justified **CL*** in this particular case but in all cases. The general argument is, as already indicated, that every single cause determines the same effect given the right circumstances; that is singular causes are not indeterministic.

The second issue we have to face is what are the relata in causal laws. In opposition to the regularity account, some realist philosophers want to argue that laws of nature are relations between universals. I have no troubles with universals. It is something similar in each instance of a group of things that makes them the same sort of thing. Things do have common properties that make it possible to name them with kind names. If they didn't, we would fail to make a unique and non-arbitrary reference to other things than particulars. If some particulars didn't share certain observable properties, we could only use a proper name, and not a common name, when referring to this or that particular. Although the common properties are recognized as 'qualitatively' identical by being abstracted from particulars, they are nonetheless numerically different, existing only in particulars as concrete instantiated properties. The individuation of universals is logically tied up with particulars. We cannot meaningfully ask whether universals are real separately from our ability to use common names in particular situations. I hold that nothing exists except actual particulars and actual or categorical properties and relations.

It is, of course, true that everything is similar to everything else in one way or another. A dog and its owner are similar to each other by living in the same house. This similarity is not shared with a stray dog and a person owing no dog. But most of the similarities, the owner and the dog share, are not real, positive properties. The owner and the dog have in common an infinite number of negative properties: by not living on the moon, etc.; by not living in the seventeenth century, etc.; by not being ten feet high, etc.; by not being made of stone, etc., etc. But apart from living in the same house and being both mammals, the dog and the owner have many more positive properties in common with their own species than with each other. The positive properties they share with their species are at least those characteristics which are necessary for us to be acquainted with in order to recognize something as a dog or as a human being.

Notice it has not been argued that things falling under distinct common names cannot be brought under the same name. Neither has it been argued that things falling under the same name cannot be split up and put under different names. My claim is only that if this happens, the things in question must share some positive attributes that make a new identification possible. Moreover, the suggestion does not imply that those things loose properties they previously had, or gain some new ones. For if the original identification remains, the various particulars must still satisfy those sortal predicates that necessarily ascertain the correct use of the previous names. There is, indeed, a contextual element involved here because we might wish to identify something differently in different situations where our cognitive interests have changed. First and foremost we get to universals by predication based on invariant perceptual clues. They are there whether we are interested or not. A certain subjective factor may nevertheless enter the partition and categorisation of the world – though not necessarily in an arbitrary way. In general we group things together under a common name if we believe that they demarcate themselves from other things by possessing more similar observable, non?relational, properties of some kind than dissimilar ones. Usually, we take this to be the correct way of naming things. But sometimes we do put a common name on things because, from a certain viewpoint, they share some interesting property. Hence in the latter cases we cannot say that one way of dividing things is more correct than another way.

The realist's suggestion is that laws of nature are necessities or relations between certain entities. But what kind of necessity, or what kind of relation, is a law? Here contrasting opinions have been aired. Some realists, such as A. C. Erwing, C. Swoyer, and S. Shoemaker have argued in favour of a necessitarian account of laws in terms of logical necessity. Others, like Kneale and Popper, go for a necessitarian account in terms of natural necessity; and again others, like *F*. Dretske, M. Tooley, and D. Armstrong, see laws as a contingent relation between universals. All three accounts miss the fundamental distinction between theoretical laws and causal laws; therefore they inevitably fail.

The first suggestion maintains that laws of nature hold in every possible world. But this is simply not true for causal laws. Usually they are valid only for a very restricted subset of possible worlds, namely those that satisfy the nomologically relevant conditions. Even the actual universe is not a place in which the same causal laws are realized at every place and at every time in its history. Only theoretical laws are claimed to be true by stipulation in every possible world.

The second suggestion takes natural laws to be valid in every naturally possible world. Apart from being a subject to the same objections as the first account, it generates its own problems. Although Popper takes the laws of nature to be naturally or physically necessary, he also says: "A statement may be said to be naturally or physically necessary if, and only if, it is deducible from a statement function which is satisfied in all worlds that differ from our world, if at all, only with respect to initial conditions." (Popper, 1959, p. 433) But natural necessity is, in general, explicated in terms of naturally possible worlds which again are defined as the subset of worlds having all and only the same laws of nature as the actual world. How then can we explain the nature of laws of nature by an appeal to the very concept of natural laws? Neither Popper nor anybody else has ever given a satisfactory analysis avoiding such circularity.

This leads us to the third suggestion. Like the second account, it holds that the nomical states of affairs are physically necessary. But the states of affairs are relations, and the law is taken to be more than a collection of single cases of necessitation. Assuming that it is a law that Fs are Gs, then it is equivalent to say, according to Armstrong, that "*being an* F *necessitates being a* G and, *because* of this, each individual F must be a G." (Armstrong, 1983, p. 78) Things and events belong to the same class of objects because they have a common essence, and it is the relationship between these essences that make up the foundation of a law. Thus, the nomic relation holds in virtue of something essentially an F physically determines something different to be essentially a G. In addition to the particular problem of regarding all laws as relations, the essentialist account fails for some of the same reasons as the other two.

It is not true that every law is a *relation* between universals, since, apparently, there are fundamental laws that do not relate anything at all. If we for the sake of argument consider exclusion laws like "Nothing moves with a speed greater than the velocity of light" and "No perpetuum mobile of the second order exists" as genuine laws of nature, then we cannot say that these laws contain a relation. Conservation laws are another group of laws which are not expressible in terms of relations. A law like "In every closed and isolated system the energy is preserved" may hold by necessity of some sort, but the necessity does not concern a relation between universals.

But even if we accept laws as relations it will not help Armstrong. On the one hand, if we read the word 'necessitates' as expressing a causal connection, then it misfires because there is no such causal necessitation without exceptions. So no such connection between essential properties exists. On the other hand, if we take the equivalent formulation to express a theoretical law, i.e. a relation between quantities, then the word 'necessitates' does not signify a *de re* modality. In contrast, as I shall argue, the word expresses a *de dicto* modality, stating the sortal necessity between being an F and a G.

All three accounts rest on the idea that there is some kind of necessity, which permits laws of nature to differ from their actual manifestations. The account being defended here takes another stand. Causal laws also go beyond the mere uniformity and association, but causal necessitation is not something that can lie beyond any possible experience, because causes are grounded in the way we see nature. Likewise, theoretical laws cannot be associated with regularities or uniform successions. They express necessities too, but by having no descriptive content they have in some sense more in common with legal laws than with causal laws.

Acknowledgements. An early version of this paper was presented as "Laws of Nature" at Centre National de la Recherche Scientifique, Paris, France, 23 January 1996. I want to thank Catherine Chevalley for her kind invitation. I also want to thank Paul Needham and Igor Hanzel for their helpful comments.

References

Aristotle (1939). *De Caelo.* Trans. *Aristotle on the Heavens.* Harvard University Press, Cambridge Mass.

Armstrong, D. (1983). *What is a Law of Nature?* Cambridge University Press, Cambridge.

Cartwright, N. (1983). *How the Law of Physics Lie.* Clarendon, Oxford.

Ellis, B. (1963). Universal and differential forces. *The British Journal for the Philosophy of Science*, 14:177–194.

Ellis, B. (1965). The origin and nature of Newton's laws of motion. In Colodny, R. G., editor, *Beyond the Edge of Certainty*, Englewood Cliffs. Prentice-Hall.

Faye, J. (1989). *The reality of the future.* Odense University Press, Odense.

Faye, J. (1994). Causal beliefs and their justification. In J.Faye, Scheffler, U., and Urchs, M., editors, *Logic and Causal Reasoning*, pages 141–168, Berlin. Akademie Verlag.

Faye, J. (2000). Observing the unobservable. In Agazzi, E. and Pauri, M., editors, *The Reality of the Unobservable Observability, Unobservability and their Impact on the Issue of Scientific Realism*, pages 165–177, Dordrecht/Boston/London. Kluwer Academic Publishers.

Faye, J. (2002). *Rethinking Science.* Ashgate, Aldershot.

Giere, R. (1988). *Explaining Science.* The University of Chicago Press, Chicago.

Hanson, N. R. (1958). *Patterns of Discovery.* Cambridge University Press, Cambridge.

Kuhn, T. B. (1962). *The Structure of the Scientific Revolutions.* The University of Chicago Press, Chicago.

Newton, I. (1946). *Sir Isaac Newton's Mathematical Principles of Natural Philosophy and his System of the World.* University of California Press, Berkeley. (Translation A. Motte, revised by F. Cajori).

Poincaré, H. (1952). *Science and Hypothesis.* Dover, New York.

Popper, K. R. (1959). *The Logic of Scientific Discovery.* Hutchinson, London.

Ramsey, F. P. (1978). Law and causality. In Mellor, D. H., editor, *Foundations*, pages 128–151, London. Routledge & Kegan Paul.

Sklar, L. (1982). Saving the noumena. *Philosophical Topics*, 13:49–72.

Tooley, M. (1997). The nature of law. *Canadian Journal of Philosophy*, 7:667–698.

Toulmin, S. (1953). *The Philosophy of Science.* Hutchinson, London.

NANCY CARTWRIGHT AND LESZEK NOWAK ON SCIENTIFIC LAWS AND SCIENTIFIC EXPLANATION

Igor Hanzel

The aim of this paper is to give an analysis of Nancy Cartwright's views as they developed in the last twenty years or so. I start with an analysis covering her views from the end of the seventies till the mid eighties (the so-called *Lies*-period), and then I deal with her views from the late eighties till now (the so-called *Capacities*-period). What follows then is, first, a comparison of Cartwright's views from the *Lies*-period with the views of L. Nowak and, second, a comparison of the latter with her views from the *Capacities*-period. It is a contention of this paper that her views from the *Lies*-period are the result of the fact that she did not draw upon the *positive* aspects of L. Nowak's works (Nowak, 1972; Nowak, 1980) simply because she did not know these works at all at that time. On the other hand, her views from the *Capacity*-period, even if she knew at that time the works of L. Nowak, are the result of the fact that she did not realize the restrictions inherent to Nowak's (Nowak, 1972; Nowak, 1980).

1. N. Cartwright on Laws and Scientific Explanation

Nancy Cartwright's works (Cartwright, 1979 through Cartwright, 1999), contrary to the previous orientation of the philosophy of science exclusively on theoretical science, concentrate on the laboratory work of the applied physicist and the practices of the engineer. Still, in my view, she faces profound problems when dealing with theoretical science (especially theoretical physics), and this holds not only for her Lies-period but for the *Capacities*-period as well. I start with the former.

J. Faye, P. Needham, U. Scheffler and M. Urchs, (eds.), Nature's Principles, 103–135.

1.1 Laws as Lies

The principle idea of the *Lies*-period can be summarized as a *whole-scale attack on fundamental, high level, abstract laws of science.* These laws (Cartwright, 1980a, 75; Cartwright, 1983, 55)

> do not describe true facts about reality. The fundamental laws of physics [...] do not tell what the objects do in their domain. If we try to think of them in this way, they are simply false, not only false, but deemed false by the very theory which contains them.

In opposition to fundamental laws, she prefers, as an empirically minded philosopher of science, phenomenological laws. What is real is, in her view, concrete, while our scientific abstraction of the real in fundamental (abstract) theories gives only a distorted view of the real. As I will show below, her main argument against the truth of fundamental (abstract) laws of science is that they contain idealizations (*ceteris paribus* clauses). Even if, in general, idealizations, she claims, are not harmful to science (Cartwright, 1983, 111), they represent a problem when one deals with the relation of fundamental laws to phenomenological laws and – as I will show below – especially when one tries to understand this relation by means of Hempel's D-N model of scientific explanation. Because she rejects this model, she speaks out against laws but in favour of causes. "[T]he real content of our theories is in the detailed causal knowledge they provide of concrete processes in real materials," (Cartwright, 1983, 128) and where this kind of knowledge is expressed not in fundamental (abstract) laws but in the true "low level causal principles and concrete phenomenological laws" (Cartwright, 1983, 10).

Now what does N. Cartwright state about scientific explanation? She mentions two cases of *ceteris paribus* laws that should prove that fundamental laws of physics lie. First, she mentions the law of gravity and Coulomb's law (Cartwright, 1980a, 76-77; Cartwright, 1983, 57). The force of interaction between two objects is according to the former law given by means of the equation $F = Gmm'/r^2$. But if one asks if the law of gravitation truly describes the behaviour of objects, then the answer is definitely *no.* Namely (Cartwright, 1980a, 77; Cartwright, 1983, 57)

> [i]t is not true that for any two bodies the force between them is given by the law of gravitation. Some bodies are charged bodies, and the force between them is not $F = Gmm'/r^2$ [...] No charged objects will behave just as the law of universal gravitation says; and any massive objects will constitute a counterexample to Coulomb's law. These two laws are not true, they are not even approximately true.

Of course, if one adds the *ceteris paribus* clause that the objects are not charged and that non-gravitational forces do not act, then the law of gravitation is true. But then (Cartwright, 1980a, 77; Cartwright, 1983, 58)

> it is not a very useful law. One of the chief jobs of the law of gravity is to help explain the forces which objects experience in various complex circumstances. This law can explain in only very simple, or ideal circumstances. It can account for why force is as it is when just gravity is at work, but it is of no help for cases in which both gravity and electricity matter. Once the *ceteris paribus* modifier has been attached, the law of gravity is irrelevant to the more complex and interesting situations.

As the second case of a *ceteris paribus* situation she mentions the changes of the energy levels of a carbon atom due to various intervening causes (Cartwright, 1980a, 80–81; Cartwright, 1983, 67–69). Initially, if the difference between the exact Coulomb interaction and the average potential is zero ($V_1 = 0$) and the spin-orbit coupling is put equal to zero as well ($V_2 = 0$), then one obtains only one energy level for that type of atom. But if one supposes that only the spin-orbit coupling is zero, but $V_1 \neq 0$ holds, then one obtains already three energy levels. Finally, by abolishing the latter *ceteris paribus* condition (i.e. $V_1 \neq 0$ and $V_2 \neq 0$ holds), one obtains a total of five energy levels. The moral, which N. Cartwright draws from these two examples, is that in the case of fundamental laws we face an *unbridgeable gap between their factual content and their explanatory power.* Either they are true, in a certain domain delineated by the *ceteris paribus* clauses, but then they cannot serve as a basis of explanation, because their application is limited to that domain, or, if one gets rid of the *ceteris paribus* clauses and adjusts the fundamental law to the conditions prevailing outside that domain, so that they can serve for the purpose of explanation, they loose their truth. So, she claims, in fundamental laws there is "a trade-off between factual content and explanatory power" (Cartwright, 1980a, 83; Cartwright, 1983, 72).

That the fundamental laws of physics are not true is, according to her, also readily seen when one investigates more closely into the process of their application to real experimental and technological practices. As mentioned above, in order to explain on the basis of a fundamental *ceteris paribus* law one has to get rid of the *ceteris paribus* clauses, i.e., one has to improve on them, give them *factual content*, change them into *phenomenological laws.* This means that fundamental laws are by themselves empty, devoid of any cognitive content. She, therefore, also claims

> [t]he great explanatory and predictive powers of our theories lies in their fundamental laws. Nevertheless the content of our scientific knowledge is expressed in the phenomenological laws (Cartwright, 1983, 100).

Cartwright distinguishes two types of improvements on the equations of abstract laws. The first, the so-called *ab vero*, she illustrates by the following example (Cartwright and Nordby, 1983, 274–275; Cartwright, 1983, 105–106). The speed of an airplane is calculated according to the following equation

$$V_E = [2(P_T - P_0/r_S) \times 1/(1 + M^2/4 + M^4/40 + M^6/1600 \ldots)]^{\frac{1}{2}} \quad (1)$$

where the terms in the denominator are members of a Taylor series and M stands for Mach's number. If for the speed of an airplane holds $M < 1$, then it is possible to discard all but the first three terms in the denominator, and one obtains

$$V_E = [2(P_T - P_0/r_S) \times 1/(1 + M^2/4 + M^4/40)]^{\frac{1}{2}} \quad (2)$$

Finally, in the case when for the speed of an airplane holds $M < 0.5$, one obtains for its speed

$$V_E = [2(P_T - P_0/r_S) \times 1/(1 + M^2/4)]^{\frac{1}{2}} \quad (3)$$

The second of improvements of an abstract equation, the so-called *ad verum improvement*, is illustrated by the following example (Cartwright and Nordby, 1983, 276–277; Cartwright, 1983, 107–110). If one constructs an amplifier, then he or she calculates initially its midband gain by means of the following formula

$$A_\nu = R_L/[kT/qI_E + (r_b + R_S)(1 - \alpha)] \quad (4)$$

But calculations based on it yield a result which differs significantly from what is measured on a real amplifier This discrepancy is explained in such a way that 4 does not take into account a certain undiagnosed combination of causal factors which requires to replace the term kT/qI_E by the term $(k + k_0)T/qI_E$. 4 does not take into account also the series resistance because real electrolytic capacitors are not ideal. One has, therefore, to introduce the additional term r_{c1} into the denominator. These two modifications change 4 into

$$A_\nu = R_L/[(k + k_0)T/qI_E + (r_b + R_S)(1 - \alpha) + r_{c1}] \quad (5)$$

Calculations based on this equation give values, which are quite close to those actually measured.

This second type of an improvement of an abstract equation from a fundamental law again, according to Cartwright, shows that the defenders of fundamental (abstract) laws have it wrong because (Cartwright, 1983, 111)

> the improvements come at the wrong place for the defenders of fundamental laws. They come from the ground, so-to-speak, and not from the top down. We do not modify the treatment by deriving from our theoretical principles a new starting equation to replace [4 ...] What we do instead is to add a phenomenological correction factor, a factor that helps produce a correct description, but that is not dictated by fundamental law.

Because fundamental laws are used for the explanation of phenomenological laws, but still are, contrary, to the latter not true, she proposes a simulacrum account of explanation, so that instead of understanding scientific explanation as a process going from fundamental laws directly to phenomenological laws, explanation proceeds from fundamental laws via a model to phenomenological laws, so that "[t]he phenomenological laws are indeed true of the objects in reality [...] the fundamental laws are true only of objects in the model" (Cartwright, 1983, 4). Drawing upon the *English Oxford Dictionary* she states (Cartwright, 1983, 17)

> a simulacrum is 'something having merely the form or appearance of a certain thing, without possessing its substance or proper qualities.' On the simulacrum account, to explain a phenomenon is to construct a model which fits the phenomenon into a theory. The fundamental laws of the theory are true of the objects in the model and they are used to derive a specific account of how these objects behave. But the objects of the model have only 'the form or appearance of things' and in a very strong sense, not their 'substance or proper qualities' [...] The lesson for the truth of fundamental laws is clear: fundamental laws do not govern objects in reality; they govern only objects in models.

The simulacrum account of explanation is viewed by Cartwright as "an alternative to the D-N model that brings the philosophical account closer to explanatory practices in physics" (Cartwright, 1983, 151). In my view Cartwright displays in the *Lies*-period an ambivalent approach to the D-N model. On the one hand, she criticizes it because, as her deliberations on explanation in physics show (Cartwright, 1980b, 159; Cartwright, 1983, 45),

> [w]e cannot explain [...] phenomena with a covering-law model [...] because we do not have laws that cover them. Covering laws are scarce. Many phenomena which have perfectly good scientific explanations are not covered by any laws. No true laws, that is. They are at best covered by *ceteris paribus* generalizations – generalizations that hold only under special conditions, usually ideal conditions.

She criticizes the D-N model also because "it is patently mistaken when on looks at real derivations in physics or engineering. It is never strict deduction that takes you from the fundamental equations at the beginning to the phenomenological laws at the end" (Cartwright, 1983, 104).

But on the other hand, she often displays a believe into certain proclaimed characteristics which, as I will show below, are simply not true of it. So, e.g., after putting the question "[h]ow do we fit a phenomenon into a general theoretical framework?" (Cartwright, 1983, 16) she states (Cartwright, 1983, 16)

> [p]rima facie, the covering-law model seems ideally suited to answer: we fit a phenomenon into a theory by showing how various phenomenological laws which are true of it derive from the theory's basic laws and equations [...] The 'covering' of 'covering-law model' is a powerful metaphor. It teaches [...] that phenomenological laws can be derived from fundamental laws.

The fact that the D-N model had such an profound impact on Cartwright's understanding of the process of explanation, that she was not able in the *Lies*-period to provide an alternative to it, is readily seen in her account of the process of explanation of the changes of the energy levels of the carbon atom given above. (Cartwright, 1980a, 81; Cartwright, 1983, 69)

> [i]t is hard to state a true factual claim about the effects of the Coulomb potential in the carbon atom. But quantum theory does guarantee that a certain *counterfactual* is true; the Coulomb potential, if it were the only potential at work, would produce the three [energy] levels [...]. Clearly this counterfactual bears on our explanation. But we have no model of explanation that shows how. The covering-law model shows how statements of fact are relevant to explaining a phenomenon. But how is a truth about energy levels, which could occur in quite different circumstances, relevant to the levels which do occur in these? We think the counterfactual is important; but we have no account of how it works.

So, Cartwright in this case does not know how to reconstruct the explanation of changes of the number of energy levels, where this explanation starts from the state where $V_1 = 0$ and $V_2 = 0$ holds, and proceeds via $V_1 \neq 0$ and $V_2 = 0$, finally to arrive at the explanation when $V_1 \neq 0$ and $V_2 \neq 0$ holds. Because she was not aware of the process of gradual concretization, reconstructed by Nowak in (Nowak, 1972), she claimed eleven years after the publication of Nowak's paper also that "a law that holds only in restricted circumstances can explain only in those circumstances" (Cartwright, 1983, 155).

1.2 Laws as Capacity Claims

The *Lies*-period ends with N. Cartwright claiming that according to the simulacrum account (Cartwright, 1983, 161)

> within the model we 'derive' various laws which match more or less well with bits of phenomenological behaviour. But even inside the model , derivation is not what the D-N account would have it to be, and I do not have any clear alternative.

The reason for this she sees in the fact that (Cartwright, 1983, 161–162)

> I do not know how to treat causality. The best theoretical treatments get right a significant number of phenomenological laws. But they must also tell the right causal stories. [...] But what is for a theoretical treatment to 'tell' a causal story? [...] I do not have a philosophical theory how it is done [...] We [philosophers of science] need a theory of explanation which shows the relationship between causal processes and the fundamental laws we use to study them, and neither my simulacrum account nor the traditional covering-law account are of much help.

The *Capacities*-period brings in respect to that the right remedies. The fundamental (abstract) laws, N. Cartwright now claims, state the underlying, inherent causal capacities of certain entities, where these capacities, together with the conditions in which they are exercised, determine the actual (phenomenological) behaviour of these entities, while the relation of the fundamental law to the concrete phenomenological laws is, she declares, that of gradual concretization as reconstructed by L. Nowak in (Nowak, 1980).

While in the *Lies*-period she stated that in respect to what fundamental laws of science claim "no story I know about causal powers makes a good start" (Cartwright, 1983, 62), the situation changes profoundly in the *Capacities*-period when she states "[c]ausal laws are best [...] rendered as capacity ascriptions [...] when capacity ascriptions play this role, they are functioning as a material abstractions" (Cartwright, 1989a, 355). While material abstractions enable us to get rid in mind of the concrete details in the domain under scrutiny, after it is finished "we are left with a law that is meant not literally true to describe the behavior of objects in its domain, but rather [...] to reveal the underlying principles by which they operate" (Cartwright, 1989a, 354).

She presents a three-tiers model (Cartwright, 1989a, 355; Cartwright, 1989, 228)

> [a]t the bottom we have individual singular causings. At the top we have general causal claims, which I render as statements associating capacities with properties – "aspirins have the power to relieve headaches" [...] In between stands the phenomenal content of the capacity claim – a vast matrix of detailed complicated causal laws.

For her singular causings are primary (Cartwright, 1989, 2–3)

> [i]t is the singular fact that matters to the causal law because that is what causal laws are about. The generic causal claims are [...] ascriptions of capacities, capacities that make things happen, case by case. 'Aspirins relieve headaches' [...] says that aspirins have the capacity to relieve headaches, a relatively enduring and stable capacity that they carry with them from situation to situation [...] which is just as surely seen in good single case.

The priority of singular causings is readily seen when one deals with the following two issues: How we can infer causes from data? How we can infer causes from theory? Her answer to both problems is unambiguous: we need some antecedent causal knowledge; singular causes in – singular causes out, and, no singular causes in – no singular causes out. She states (Cartwright, 1989, 94–95)

> [w]ithout antecedent information it is no more possible to establish a causal claim via a regularity than it is to demonstrate a singular cause directly, and in both cases the inputs must include causal information – not only information about general causal laws, but about singular facts as well [...]. Singular claims are not just input for inferring causal laws; they are the output as well.

A generic causal claim (causal law), in respect to a singular causal claim "attributes to the featured characteristic – say, being an aspirin – a capacity to produce the appropriate effect in individual cases" (Cartwright, 1989a, 350). "Generic causal laws record [...] capacities. To assert the causal law that aspirins relieve headaches is to claim that aspirins, by virtue of being aspirins, have the capacity to make headaches disappear" (Cartwright, 1989, 136). Because "[t]he cause necessitates its effect – it makes it happen or brings it about; and the occurrence of the effect is explained by the occurrence of the cause," she claims that "laws [...] describe what causes are capable of doing" (Cartwright, 1993b, 428–429). But still, somehow echoing here views from the *Lies*-period, she claims that (Cartwright, 1989, 8)

> laws are a poor stopping point. It is hard to find them in nature and we are always having to make excuses for them: why they have exceptions [...]; why they only work for models in head; why it takes an engineer with a special knowledge of real material and a not too literal mind to apply physics to reality. The point [...] is to argue that we must admit capacities, and my hope is that once we have them we can do away with laws. Capacities do more for us at a smaller metaphysical price.

Cartwright's reflection on capacities are related to the following criterion of causality, *CC*. Let C and E stand for a putative cause and its effect, while $F_1, \ldots, F_n$ designate E's other causes, and where $+F_i$ denotes that F_i is at work, while $-F_i$ denotes that it is not. She suggests then (P stands for probability)

$$CC\colon \quad C \text{ causes } E \text{ iff} \qquad (6)$$
$$P(E \mid C \pm F_1, \ldots, \pm F_n) > P(E \mid \neg C \pm F_1, \ldots, \pm F_n)$$

where $\{F_1, \ldots, F_n, C\}$ is a complete causal set of E (i.e., it includes all of E's causes). She emphasizes that 6 holds universally; there is a "universal quantifier in front: C is to increase (or at least not decrease) the probability of E in every homogeneous background [...it] quantifies over *all* test situations" (Cartwright, 1989, 143–145). It is independent of any population T; it does not state "'Cs causes Es in T', but rather 'Cs causes Es', simpliciter" (Cartwright, 1989, 145). While the former, localized claim has the character of a casual law, the delocalized claim is a claim about C's capacity. (Cartwright, 1989, 145)

> [i]t reports that Cs - by virtue of being C - can cause E [...] C carries the capacity to produce E [...]. If Cs ever succeed in causing Es (by virtue of being C), it must be because they have the capacity to do so. That capacity is something they can be expected to carry with them from situation to situation.

Stated otherwise (Cartwright, 1995d, 177)

> [w]e perform an experiment in certain specific circumstances; the experiment licenses a law-like regularity claim very crudely of the form "In I, A's do X. This law, however, covers hardly any cases. Most A's are not in I. We abstract to A's do X. This claim is not restricted to the circumstances I. Indeed it does not refer to any circumstances at all; these have been "subtracted".

The concept of capacity is in the works of N. Cartwright from the nineties closely related to that of *nature*[1] and *nomological machines* (Cartwright, 1997b, 65–66; Cartwright, 1999, 49–50). In respect to the former she claims that (Cartwright, 1995a, 277; Cartwright, 1999, 138)

> [w]e aim in science to discover the nature of things, we try to find out what powers or capacities they have and in what circumstances and in what ways these capacities can be harnessed to produce predictable behaviours.

The concept of capacity, nature and power also directly links Cartwright's works from the *Capacities*-period with to the works of L. Nowak (Nowak, 1972; Nowak, 1980). In respect to the above given claim "A's do X" she states (Cartwright, 1995d, 178)

---

[1] On this see, e.g., her (Cartwright, 1992), where she states that "our basic knowledge – knowledge of capacities – is typically about natures, and what they produce" (Cartwright, 1992, 46; Cartwright, 1999, 80).

> [w]e abstract "A's do X" from special test situations like I. We apply it to far more complex situations where varieties of factors other than A are at work [...] the abstract claim, I maintain, is the kind we assert in our theories in those domains where explanation and prediction proceed by the method of concretization [...] we use abstract claims as if they were ascriptions of tendencies (where the claim is about what things *do*) or of capacities (where the claims about what things *cause*). How do we get from an abstract claim of the form *A's do X* to a concrete regularity type law? Not by deduction of the kind pictured by Hempel, since we do not have a statement of a covering law to begin from. Nowak's alternative is the process of concretization as he describes it.

So, N. Cartwright distinguishes here *two different, but still interconnected phases of scientific cognition.* On the one hand, the process of "'going upward' from experience to general principles," and, on the other hand, the movement "'[...] downwards from that general principle to a variety of specific conclusions'," (Cartwright, 1989, 183) where the latter has the character of a concretization process. This means that she clearly distinguishes here "the *converse* process of abstraction and concretization," (Cartwright, 1989, 184) i.e., *the movement from the concrete to the abstract, and from this "back" to the concrete.* The turning point of such a movement is the knowledge of the capacity stated in the abstract (fundamental) law, which – she claims – holds only in ideal circumstances, in an ideal situation. "What is an ideal situation for studying a particular factor? It is a situation in which all other 'disturbing' factors are missing. And what is special about that? *When all other disturbances are absent, the factor manifests its power explicitly in its behaviour*" (Cartwright, 1989, 190-191).

2. L. Nowak on Laws and Scientific Explanation

When I spoke above about the "mythological" characteristics of the D-N model I meant the following one. Hempel in Part I of the study (Hempel and Oppenheim, 1948) claimed that his approach can reconstruct the case when the explanandum-sentence expresses a law. But in his attempt to provide a definition for scientific explanation by means of a simple model language in Part III of this study, he has already encountered a problem, articulated in footnote 33, which forced him to restrict his reconstruction of scientific explanation to apply only to a case of explanation of singular events (Hempel and Oppenheim, 1948, 273):

> This is not a matter of free choice: The precise rational reconstruction of explanation as applied to general regularities presents peculiar problems for which we can offer no solution at present. The core of the difficulty can be indicated briefly by reference to an example: Kepler's

> law, K, may be conjoined with Boyle's law, B, to make a stronger law, $K.B$; but derivation of K from the latter would not be considered as an explanation of the regularities stated in Kepler's laws; rather, it would be viewed as representing, in effect, a pointless 'explanation' of Kepler's laws by themselves. The derivation of Kepler's laws from Newton's laws of motion and of gravitation, on the other hand, would be recognized as a genuine explanation in terms of more comprehensive regularities, or so-called higher-level laws. The problem therefore arises of setting up clear-cut criteria for the distinction of levels of explanation or for a comparison of generalized sentences as to their comprehensiveness. The establishment of adequate criteria for this purpose is as yet an open problem.

The solution to Hempel's footnote 33 problem was provided by Leszek Nowak in (Nowak, 1972).[2] His innovative approach to scientific explanation is primarily based on a new approach to the structure of scientific laws. It can be represented symbolically as follows (Nowak, 1972; Nowak, 1980):

$$(x)[Gx\&p_1x = d_1\&\ldots\&p_kx = d_k \rightarrow F^{(k)}x = f_k(Hx)] \tag{7}$$

"G" is a predicate letter denoting the class of objects for which the law is formulated (the universe of discourse), "p_1," ..., "p_k," "H" and "$F^{(k)}$" are function terms denoting functions defined on the class, denoted as "G," over which the individual variable "x" ranges. Here "F" denotes the phenomenon to be explained, "H" denotes the factor, which is principal for the explained phenomenon; "p_1," ..., "p_k" denote secondary factors – modification conditions – that can have an impact on the explained phenomenon, they can modify it. "d_1," ..., "d_k" are names for certain numbers, "f_k" is a name for a function defined on the set of values of the function denoted by "H" and with values in the set of values of the function denoted by "F." "(k)" as an upper index denotes the number of idealizational assumptions, where the latter, according to Nowak, can be symbolized as "$p_i = d_i$" (for $i = 1, \ldots, k$), so that the numbers denoted by "d_i" are the extreme elements of that set of numbers from which the function, denoted by "p_i," takes on its values.

Nowak, after explicating his approach to the structure of scientific laws, proposes his model of scientific explanation. It is based on the

[2]W. C. Salmon claims that "this problem was not seriously attacked until more than a quarter of a century later – when Michael Friedman finally took it on," (Salmon, 1990, 94) and proves that his attempt "cannot get off the ground" (Salmon, 1990, 99). In my view Nowak not only attacked that problem two years before Friedman's (Friedman, 1974), but his approach gets off the ground. Surprisingly, W. C. Salmon does not mention Nowak's paper (Nowak, 1972) in the detailed bibliography covering four decades of philosophical investigation in scientific explanation since the publication of (Hempel and Oppenheim, 1948).

idea that "the idealizational assumptions are removed one by one, this brings the law closer to facts, and appropriate corrections, resulting from the removal of those assumptions, are introduced into the consequent of the law" (Nowak, 1972, 537). This process of explanation, based on scientific laws with a structure corresponding to 7, and labelled *scientific explanation by gradual concretization*, can be symbolically represented as follows ($1 \leq j \leq k+1$)

$$(x)[Gx \& p_k x \neq d_k \& \ldots \& p_1 x = d_1 \rightarrow F^{(k-1)}x = f_{k-1}(Hx, p_k x)] \quad (8)$$

$$(x)[Gx \& p_k x \neq d_k \& \ldots \& p_{k-j}x \neq d_k \& p_{(k-j-1)-1}x = d_{(k-j-1)-1} \rightarrow F^{(k-j)}x = f_{k-j}(Hx, p_k x, \ldots, p_{k-j}x)] \quad (9)$$

where "$p_i \neq d_i$" means that an idealization has already been abolished.

Nowak's approach, in my view, conveys, in comparison with the D-N model a richer view of scientific explanation. According to Hempel's D-N model it has the character of a deductive subsumption: *subsumption* of the particular event we want to explain under a *covering* law *plus deduction* of this event from laws and singular (initial and/or boundary) conditions. According to Nowak, scientific explanation of a particular event contains *not two but three moments*. Subsumption of the particular event we want to explain under a law, where this particular event is *not covered by this law as a whole*, but only by the universe of discourse and the principal factor stated in it. Plus a *move going in the opposite direction*, the *concretization* of the law to the modification conditions of the particular event to be explained, plus, *only at the very end*, the deduction of the particular event to be explained by the introduction of the singular into the already concretised scientific law. In accordance with L. Nowak's (Nowak, 1972, 538) let me introduce the sign "$\dashv$" for gradual concretization, "$L^{(k)}$," "$L^{(k-1)}$," "$L^{(k-j)}$" as abbreviations for 7, 8, and 9 and representing laws of the kth, $(k-1)$th and $(k-j)$th degree of idealization, while "C_{sin}" stands for singular conditions, "$\supset$" for entailment and "E" for the explanandum-event. Nowak's approach to scientific explanation can then be symbolically represented as follows:

$$L^{(k)} \dashv L^{(k-1)} \dashv L^{(k-j)} \& C_{\text{sin}} \supset E \quad (10)$$

In this scheme the move from $L^{(k)}$ to $L^{(k-1)}$, ..., $L^{(k-j)}$ represents a reconstruction of the explanation of scientific laws from scientific laws.

So, from 10 it is evident that *Nowak's model of scientific explanation by gradual concretization reconstructs, contrary to the D-N model, the case of explanation of scientific laws from scientific laws.*

3. N. CARTWRIGHT vs. L. NOWAK

L. Nowak's works (Nowak, 1972; Nowak, 1980) have, I claim, a profound impact on the evaluation of the *Lies*- and *Capacities*-period in N. Cartwright 's work.

3.1 Laws as Lies vs. Nowak

Let us first turn to the lesson N. Cartwright draws from the reconstruction of the splitting of the energy levels of the carbon atom. She states (Cartwright, 1980b, 159; Cartwright, 1983, 45–46)

> [c]*eteris paribus* generalizations, read literally – without the '*ceteris paribus*' modifier – as laws, are false [...] On the other hand, with the modifier the *ceteris paribus* generalizations may be true, but they cover only those few cases where the conditions are right. For most cases, either we have a law that purports to cover, but can't explain because it is acknowledged to be false, or we have a law that doesn't cover.

This means that the problem Cartwright faces here is to give a reconstruction of the process of explanation of certain phenomena that proceeds from a *certain ceteris paribus* law which does not cover these phenomena. L. Nowak's approach to scientific explanation gives the correct answer. When we try to explain a certain phenomenon, *we subsume it not under a law that as a whole covers it, but under a ceteris paribus (idealizational) law, whose only universe of discourse (G in 7) and its principal factor (H in 7) cover the phenomenon to be explained.* What steps in after such a process of subsumption is the process of gradual concretization of the law to the modification conditions of that phenomenon, plus (afterwards) the introduction of its singular conditions and the deduction of that phenomenon. So, e.g., in the case of the splitting of the energy levels of the carbon atom, one starts (Messiah, 1961, 701–703) from the Schrödinger law with the structure

$$L^{(2)}: \quad (x)[Ox \& V_1x = 0 \text{ and } V_2x = 0 \\ \rightarrow ih\partial\Psi x/\partial tx = H_c x\Psi x] \tag{11}$$

where "O" represents an object whose quantum of action is comparable with Planck's constant and whose speed is much smaller than that of light; $V_1 = 0$ and $V_2 = 0$ represent the already mentioned idealizations, while on the right side we have the Schrödinger equation with H_c as the Hamiltonian for the central field approximation. The solution of this equation gives one energy level. If we abolish the first idealization in 11, $L^{(2)}$ changes into

$$L^{(1)}: \quad (x)[Ox \& V_1x \neq 0 \text{ and } V_2x = 0 \\ \rightarrow ih\partial\Psi x/\partial tx = (H_cx + V_1x)\Psi x] \tag{12}$$

The introduction of the correction V_1 into the Hamiltonian expresses the difference between the exact Coulomb potential and H_c. The solution of the Schrödinger equation in 12 gives three energy levels. Finally the abolishment of the last idealization takes into account the spin-orbit coupling (while still supposing that each electron in the atom moves independently from other electrons) leads to

$$L^{(0)}: \quad (x)[Ox \& V_1x \neq 0 \text{ and } V_2x \neq 0 \rightarrow \\ ih\partial\Psi x/\partial tx = (H_cx + V_1x + V_2x)\Psi x] \tag{13}$$

The solution of the Schrödinger equation here leads to five energy levels.

So the solution to N. Cartwright's dilemma: either cover without the *ceteris paribus* modifier (but then false) or not cover, but then explain not much enough (due to the restrictions imposed by that modifier) is as follows.

> *The explanatory truth of an idealizational (ceteris paribus) law cannot be judged by its ability to cover singular phenomena, but by its ability to explain the latter by means of gradual concretization and (only then) by a covering D-N explanation.*

From Nowak's approach to explanation several consequences follow which either refute Cartwright's views from the *Lies*-period, or which solve problems she was not able to solve at that time. First, it refutes the above given claim by her that "a law that holds only in restricted circumstances can explain only in those circumstances" (Cartwright, 1983, 155). If one compares the latter claim with the above given characterization of Hempel's D-N model, then it is readily seen that it is a result of the influence of that model on Cartwright's understanding of the process of explanation. Because in the process of explanation, *according to the D-N model*, we operate in mind *only with singular conditions*, a scientific law containing *ceteris paribus* clauses, can explain only in the restricted domain delineated by those clauses. But the above given reconstruction of the explanation of energy levels shows that scientific explanation can be based on thought operations not only with singular conditions but also with modification conditions. For example, $L^{(2)}$ from 11, even if it holds in the domain where the idealizations $V_1 = 0$ and $V_2 = 0$ hold, still can be used as a basis of explanation of what is going outside this domain. So, e.g., Messiah investigates not only into the cases given above when $V_1 \neq 0$, $V_2 = 0$ and $V_1 \neq 0$, $V_2 \neq 0$ hold, but also the case when $V_2 >> V_1$, i.e., when $V_1 = 0$ and $V_2 \neq 0$ holds. The Hamiltonian for this case is $H = H_c + V_2$ while Schrödinger's law is as follows

$$L^{(1)}: \quad (x)[Ox\&V_lx = 0\&V_2 \neq 0 \\ \rightarrow hi\partial\Psi x/\partial tx = (H_c + V_2x)\Psi x] \tag{14}$$

And of course $L^{(2)}$ serves for the explanation of very different phenomena, and is still true for all of them: it serves as a basis for the explanation of *one*, *three* and *five* energy levels of a concrete type of atom.

So while N. Cartwright in the *Lies*-period attacked fundamental *ceteris paribus* laws, claiming that they lie, the opposite is true of them. *The more idealizations a scientific law contains, i.e., the more abstract it is, the more different phenomenological laws and singular phenomena it can potentially explain.* As shown above in the case of the carbon atom, just two idealizations enable us to derive in quantum mechanics a network of three phenomenological laws. I add "potentially" because *in the course of explanation by gradual explanation we have to discover the respective causal impact of the acting modification condition.* This means that the process of explanation by gradual concretization does not have the character of a deductive argument. It is not a derivation of cut-and-dried cognition in the form of consequences from given premises, but a heuristic process creating new knowledge of the causal genesis of the phenomena $F^{(k-1)}, \ldots, F^{(k-j)}$; this of course grounds Cartwright's conjecture that explanation of phenomenological laws from fundamental *ceteris paribus* (idealizational) laws does not have the character of deduction. This heuristic nature of explanation by gradual concretization enables us also to reconstruct N. Cartwright's *ad verum* and *ab vero* movements from abstract laws.[3] Let us take the case of the amplifier given above in Part 1.1. Let A stands for "amplifier" and B be the shorthand abbreviation for the equation 4. The scientific law from which the *ad vero* movement begins is

$$L: \quad (x)(Ax \rightarrow Bx) \tag{15}$$

[3]I draw here upon I. Nowak's (Nowak, 1974).

If we make the actual measurement and compare its result with that on the basis of B, we discover a big difference, i.e., B does not hold for the actual amplifier. We discover that, as shown above, that one has to take into account the constant k_0; this means that equation B in 15 holds only if $k_0 = 0$. So, 15 changes into

$$L^{(1)}: \quad (x)(Ax \& k_0 = 0 \rightarrow Bx) \tag{16}$$

16 differs from 15 by the discovery of a previously unknown (hidden) idealization. Suppose further (diverging here from N. Cartwright's example) that we want to calculate the midband gain of an amplifier on the basis of 16. We have to concretise it, and we obtain

$$L^{(0)}: \quad (x)(Ax \& k_0 x \neq 0 \rightarrow B'x) \tag{17}$$

where B' stands for $A_\nu = R_L/[(k+k_0)T/qI_E+(r_b+R_S)(1-\alpha)]$. But if we would put in the singular data of an actual circuit, the calculated result would again differ from that measured. We would later discover that neither in 17, 16 nor in 15 we have known about the leakage of current in the electrolyte, i.e., B and B' hold only if the additional idealization $r_{c1} = 0$ holds. We obtain from 16

$$L^{(2)}: \quad (x)(Ax \& k_0 x = 0 \& r_{c1} = 0 \rightarrow Bx) \tag{18}$$

Explanation by gradual concretization proceeding from 18 gives us a law by means of which we, finally, get a calculations which fits the actually measured midband gain. $L^{(1)}$ and $L^{(0)}$ obtained from 18 are

$$L^{(1)}: \quad (x)(Ax \& k_0 x \neq 0 \& r_{c1} = 0 \rightarrow B'x) \tag{19}$$
$$L^{(0)}: \quad (x)(Ax \& k_0 x \neq 0 \& r_{c1} \neq 0 \rightarrow B''x) \tag{20}$$

where B'' stands for the equation 5. The whole network of developmental relations between the laws 15 through 20 can be represented by the following scheme:

$$\begin{array}{ccccc} L_1 & \Rightarrow & L_3^{(1)} & \Rightarrow & L^{(2)} \\ & & \perp & & \perp \\ & & L_2^{(0)} & \Rightarrow & L^{(1)} \\ & & & & \perp \\ & & & & L^{(0)} \end{array} \tag{21}$$

Here "$\Rightarrow$" stands for the discovery of a previously unknown (hidden) idealization; "$\perp$" stands for the relation of gradual concretization and the lower numerical indexes indicate the order of the realization of a failure.

What has to be emphasized here is that on the one hand, because the laws of the lowest degree of idealization (here L_1, $L_2^{(0)}$, $L^{(0)}$) serve as the basis of explanation of what is going on in experiments and measurements, the realization that the concretised laws, and sometimes also the laws of the highest degree of idealization (here L_1, $L_3^{(1)}$, $L^{(2)}$) have failed, enters into the above reconstructed explanation-cum-concretization hierarchy from its "bottom." But on the other hand, the process of explanation by gradual concretization starts from the "top" of this hierarchy. For example, according to the second column of the scheme, the experimenting and measuring scientist realizes that $L_2^{(0)}$ – which was previously obtained by means of a "top-down" explanatory procedure – has failed in respect to certain measurements and then, after a close scrutiny of it, proceeds, if necessary, "up" and investigates into $L_3^{(1)}$. These two opposite directions of a thought movement are inseparably entangled. In order for the laws at the "bottom" of the concretization hierarchy to causally explain what is going on in measurement and experimentation, they have to be derived from the "higher"-level causal laws of this hierarchy. But on the other hand, in order to discover and state all the laws in this hierarchy – and this includes the discovery of previously unknown (hidden) idealizations – we need also the movement of cognition from the "bottom" to the "top" of the hierarchy. Above I claimed that the process of explanation by gradual concretization, i.e., the thought movement from the "top" to the "bottom," has the character of a heuristic process creating new knowledge of the causal genesis of phenomena. The same can be stated about the thought movement going into the opposite direction. The information about the results of measurement and experimentation enters the hierarchy at its "bottom." From there it moves gradually "up" level by level, where it is processed - by means of a discovery of hidden idealizations - in order to provide a causal understanding of the results (near failure or complete failure) of the causal explanations.

So, the moral from this *ad verum* case is that even if crucial empirical information enters at the bottom of the hierarchy 21, one cannot infer from this, as Cartwright does, that the abstract laws at the top of this hierarchy lie. The same holds for the *ab vero* case of the calculations of the speed of an airplane as given in Part 1.1 above. Let P stands for "airplane" and Q for the equation 1. The law is

$$L: \quad (x)(Px \rightarrow Qx) \tag{22}$$

If we know that the actual plane is, say, a piston engine W.W.II fighter whose speed cannot exceed $M < 1$, then we can introduce into 22 an additional idealization: let the ratio of the speed of the plane flying in a certain medium, ν_p, to the speed of sound in that medium, ν_s, be such that it holds $\nu_p/\nu_s < 1$. 22 changes into

$$L^{(1)}: \quad (x)(Px \& \nu_p x/\nu_s x < 1 \rightarrow Q'x) \tag{23}$$

where Q' stands for the equation 2. Finally, if we know that the actual plane is a plane with a wooden frame equipped with a two-stroke motorcycle engine, so that $M < 0.5$ holds, then we obtain from 23[4]

$$L^{(2)}: \quad (x)(Px \& \nu_p x/\nu_s x < 0.5 \rightarrow Q''x) \tag{24}$$

where Q' stands for the equation 3.

So while in 22 I supposed that there is no speed limit for the airplane, in 23 I already narrow down the range of its possible speed – it is less than the speed of sound – and in 24 I narrow it down even more. As a result of in such a way purportedly gradually *introduced idealizations*, I obtain also a gradually *scaled down* equation for the speed of the airplane. So, similarly to the *ad verum* case it holds here that in order to gradually scale down equation 1 and the law 22 we need both the empirical information about the actual plane (W.W.II piston engine fighter; airplane with a wooden frame, etc.) and the fundamental law's equation to scale something down and then calculate the actual speed.

3.2 Laws as Capacities vs. Nowak

Let me now investigate into N. Cartwright's understanding of the very fundamental (abstract) law and of the capacity expressed in it, which represents the turning point (or the mediating link) between the movement from the concrete to the abstract and from it "back" to the concrete.

While she uses the criterion of causality *CC* as a starting point for the understanding of fundamental (abstract) laws, Nowak's reconstruction of the structure of $L^{(k)}$ in 7 can be used also for this end. Nowak's 7,

[4]Here I suppose that the idealization from 23 $\nu_p/\nu_s < 1$ is included in the idealization $\nu_p/\nu_s < 0.5$, therefore I suppose the law 24 contains two idealizations.

in my view at least, serves better as the basis for the understanding of a cause C's capacity to produce the effect E, than Cartwright's CC. About the latter she states (Cartwright, 1989, 95–96)

> [f]ormula CC says that, for a generic causal claim to hold, the putative cause C must increase the probability of the effect E in every population that is homogeneous with respect to E's other causes. But this condition is too strong, for it holds fixed too much. The other factors relevant to E should be held fixed only in individuals for whom they are not caused by C itself. The simplest examples have the structure of Fig. 3.1.

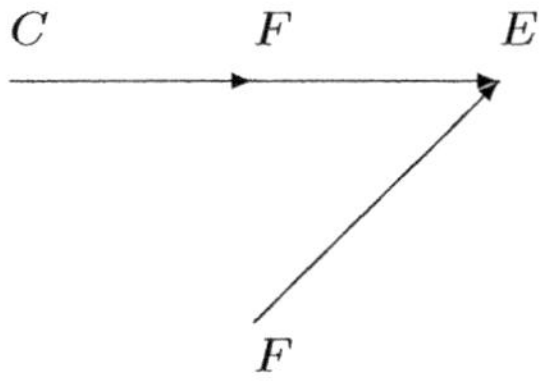

Fig. 3.1

> This is a case of a genuine cause C, which always operates through some intermediate cause, F. But F can also occur on its own, and if it does so, it is still positively relevant for E. Holding F fixed leads to the mistaken conclusion that C does not cause E. For $P(E|C \pm F) = P(E|\neg C \pm F)$. This is a familiar point: intermediate causes in a process (here F) screen off the initial cause (C) from the final outcome (E). If intermediates are held fixed, causes will not be identified as genuine even if when they are. On the other hand, if factors like F are not held fixed when they occur for independent reason, the opposite problem arises, and mere correlates may get counted as causes.

Now if we change CC so that incorporates Nowak's views expressed in 7, so that Cmod denotes a modification condition, expressed by him as "p_i," while C is the cause (principal factor) expressed by him as "H," while E is the explained phenomenon, expressed by him as "F," we obtain

CC: C causes E iff

$$P(E^{(k)} \mid C \pm C\text{mod}_1, \ldots, \pm C\text{mod}_k) > P(E^{(k)} \mid \neg C \pm C\text{mod}_1, \ldots, \pm C\text{mod}_n)$$

where $\{C\}$ is a complete causal set of $E^{(k)}$, while $+C\text{mod}_i$ means the same as $C\text{mod}_i \neq 0$ and $-C\text{mod}_i$ means the same as $C\text{mod}_i = 0$. CC in its "Nowakized" form states that C always manifests itself as $E^{(k)}$, and the latter can never be generated by any of the acting conditions $C\text{mod}_1, \ldots, C\text{mod}_k$. Of course, if for any of the modification conditions holds $C\text{mod}_i \neq 0$, then this condition together with C will produce the phenomenon $E^{(k-i)}$. $E^{(k)}$ as a manifestation of C's capacity is independent from the action or non-action of $C\text{mod}_1, \ldots, C\text{mod}_k$. So, we have

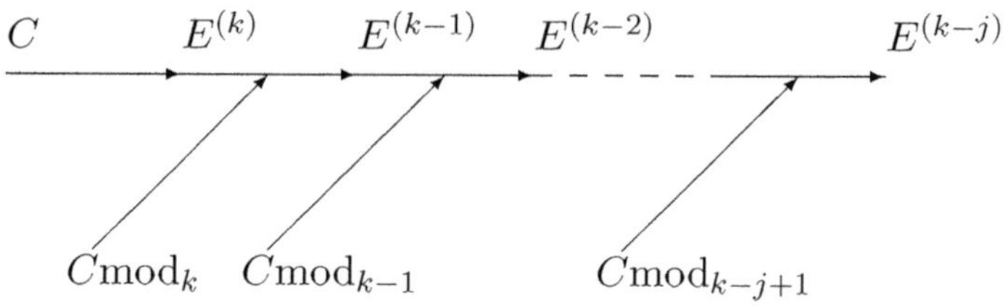

(25)

This means that without C (and its capacity to generate $E^{(k)}$), neither $E^{(k-1)}$, ..., $E^{(k-j)}$ would obtain, even if $C\text{mod}_k, \ldots, C\text{mod}_{k-j}$ would operate one by one or together. This is in fact expressed in Nowak's representation of the equation of the concretised law $L^{(k-j)}$. In our notation

$$E^{(k-j)} = f_{k-j}(C, C\text{mod}_k, \ldots, C\text{mod}_{k-j})$$

Because of C's capacity to generate $E^{(k)}$, C has always to be present in all the explanations of the various phenomena $E^{(k-1)}, \ldots, E^{(k-j)}$. But no knowledge of the causal impact of $C\text{mod}_1, \ldots, C\text{mod}_k$ is required for understanding why C manifests itself as $E^{(k)}$. This means also that in order to understand why C manifests itself as $E^{(k)}$ whatever condition $C\text{mod}_i$ is or is not at work, we have to know *in advance* C's capacity to do so. As N. Cartwright puts it "capacities are not to be identified with any particular manifestations" (Cartwright, 1997b, 75). Now the importance of Nowak's reconstruction of the structure of the equation $E^{(k)} = f_k(C)$ is as follows. Because he holds to such an understanding of the relation between a cause and its manifestation where we proceed in cognition from C to $E^{(k)}$ (i.e., we know about $E^{(k)}$ on the basis of the knowledge of C), we have to know what C's structural capacity, S, is, so that due to the latter, C manifests itself as $E^{(k)}$. Expressed otherwise, in order to write $E^{(k)} = f_k(C)$, where we proceed from the right side of this equation to its left side, we have to know *in advance* the relation of C to its structural capacity S; i.e., before we state $E^{(k)} = f_k(C)$, we have to know already the relation which can be expressed as $C = f_0(S)$. In such a way, by substituting the equation $C = f_0(S)$ into $E^{(k)} = f_k(C)$, one obtains $E^{(k)} = f_k(f_0(S))$, so that if the compound function $f_k(f_0())$ is written as $g_k()$, he or she ends up with $E^{(k)} = g_k(S)$, while 25 changes into

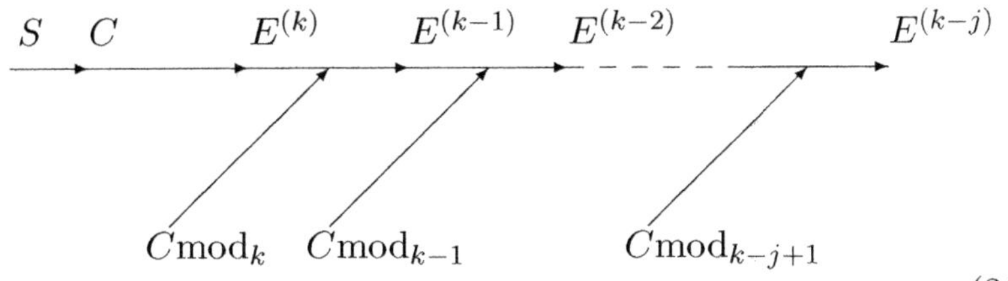

(26)

But when one looks into Nowak's (Nowak, 1980), one will find out that even if he holds to $E^{(k)} = f_k(C)$, he *never* investigates into the structure of a law containing an equation with the structure of $C = f_0(S)$ upon which the first equation *completely* depends. This is the *first negative* aspect of L. Nowak's approach in (Nowak, 1980) never realized by N. Cartwright.

So what one has to do if he or she holds to L. Nowak's $E^{(k)} = f_k(C)$ is to investigate into the fundamental (abstract) laws of science, e.g. from theoretical physics, in order to find out what specific type of capacities are stated in them. But when looks into the works of Cartwright, one will find out that she has *not provided a single detailed analysis of any fundamental law of theoretical physics to show what kind of capacity claims does it make.* Such a neglect is rather understandable in the *Lies*-period where she put emphasize on phenomenological laws used by the practical scientist and the engineer. But in the *Capacities*-period she should have hold to her own claim on the very first page of the introduction to (Cartwright, 1989)

> [t]o learn what Newtonian physics or Maxwell's electrodynamics teaches, we have to turn to what these theories say. To answer 'Are there causal powers in Maxwell's electrodynamic world?' we have studied Maxwell's laws to see if they describe causal powers or not. This is in part what I will do.

Surprisingly, nowhere in (Cartwright, 1989) or in any later works (Cartwright, 1990) through (Cartwright, 1999) of her one can find such a study into Newton's fundamental laws or Maxwell's electromagnetic theory.

Such a lack is also readily seen in her approach to quantum mechanics. According to the latter there exist three different types of changes of the state of a quantum mechanical object (i.e., of an object whose quantum of action is comparable with Planck's constant and whose speed is much smaller than the speed of light). First, the change due to its inherent dynamics; second, due to its interaction with other such objects or with radiation. Finally, changes which take place in the course of its interaction with a macroscopic object (measuring instrument). From the point of view of Cartwright's claims about capacity the first case is the most important, while the third case the least one. Even if she claims that the process of explanation based on Schrödinger equation has the character of gradual changes of the Hamiltonian, where this explanation starts from the Hamiltonian H_0 (Cartwright, 1989, 205), she never investigates into the most abstract (fundamental) form of the Schrödinger equation $ih\partial\Psi/\partial t = H_0\Psi$ in order to find out what type of capacity does it state. H_0 expresses here the Hamiltonian of a free particle (Cartwright and Nordby, 1983, 136) and the Schrödinger equation just given above expresses the development of its state (given by Ψ) *due to its own inherent dynamics.* But because the Hamiltonian H_0 is present in the Hamiltonian for any quantum mechanical object, the question N. Cartwright should have posed and tried to answer, by investigating into quantum mechanics, should have been: *What does this theory, if it does, state about the inherent capacity of any object, whose quantum of action is comparable with Planck's constant and whose speed is much smaller than the speed of light, to change its own state in time?* And Nancy Cartwright's approach to this, from the point of the *Capacity*-period, crucial question? She never puts it, and deals instead in (Cartwright, 1989) with the problem of Bell's inequalities related to measurement in quantum mechanics.[5]

Now while the claim that Schrödinger equation $ih\partial\Psi/\partial t = H_0\Psi$ by its structure corresponds to $E^{(k)} = f_k(C)$ needs further investigation in order to prove that quantum mechanics really states a capacity inherent to quantum mechanical objects, this definitely does *not* hold – as it is claimed by Nowak in (Nowak, 1980) – for the fundamental laws of classical mechanics, and especially not for Newton's second law. This is the *second negative aspect* of his (Nowak, 1980) never realized by N. Cartwright. To understand this fact let us take the second law of dynamics. The equation $F = ma$ holds rigorously only if at least two idealizations hold: the accelerated object has a negligible volume, and no forces are acting upon the physical system where this object is located. The second dynamic law of classical mechanics, therefore, appears in the form

[5]In a class on philosophy of quantum mechanics at the University of Pittsburgh in Fall 1992 I asked J. Earman if he knew about a philosophical paper dealing not with the problem of measurement in quantum mechanics but with the Schrödinger equation. His answer was negative and the situation has not, to best of my knowledge, changed since then. There does not exist a single philosophical work which would investigate into what quantum mechanics states about the inherent dynamics of its objects expressed by the Schrödinger equation of the form $ih\partial\Psi/\partial t = H_0\Psi$, where H_0 is the Hamiltonian of a free particle.

$$L^{(2)}: \quad (x)(Ox\&Id_{1,2}x \rightarrow Fx = mxa^{(2)}x) \qquad (27)$$

where O stands for an object with a non-zero mass occurring in a certain physical system, and $Id_{1,2}$ stands for the above-mentioned two idealizations.

From 27 it is readily seen that its structure does not correspond to that of 7 but to

$$(x)[Gx\&C\text{mod}_1x = d_1\&\ldots\&C\text{mod}_kx = d_k \rightarrow f_1(Cx) = E^{(k)}x] \quad (28)$$

In Newton's second law the cause C has to be grasped through the phenomenon-effect $E^{(k)}$, but in Nowak's 7 it is done the other way round. Someone could object to my distinction between 7 and 28 by suggesting that we should flip the sides of the equation $f_1(C) = E^{(k)}$ in 28 to bring it into accord with the equation (in my notation) $E^{(k)} = f_k(C)$ in 7. Let us try to do such a flipping in the equation $F = ma^{(2)}$ from Newton's second law 27. We would obtain $a^{(2)} = F/m$, but the problem here is that in classical mechanics F cannot be expressed independently from m and $a^{(2)}$. Why? *Simply because classical mechanics is a science neither dealing with the production, origin of force, nor about an entity's capacity to produce effects like change of motion, but only a science dealing with the quantity of forces and the effects of forces.* So what can be put in after that flipping into the place of F is only $ma^{(2)}$ and our flipping ends up in the triviality $a^{(2)} = a^{(2)}$.

By comparing $f_1(C) = E^{(k)}$ from 28 with $E^{(k)} = f_k(C)$ from 7 I am led to a conceptual distinction never accomplished by L. Nowak and N. Cartwright. Phenomena-effects obtained, cognised in the movement from the concrete to the abstract *before* the cognition of their cause, and where, as I will show below, *by means of the quantitative determinations of the former we determine the quantitative determinations of the latter, I label appearances of the cause.* Phenomena-effects cognised, derived *after* and *inferred from the already cognised quantity and quality of the cause and its capacity* (power) I label *manifestations of the cause and its capacity (power).* In the former, contrary to the latter, as I will show also below, the quality and capacity of the cause remains completely unknown. So even if in 7 and 28 there appears one and the same symbol $E^{(k)}$, in the former it stands for an *appearance* in the kth degree of idealisation while in the latter for a *manifestation* in the kth degree of idealisation; *their respective cognitive status in $f_1(C) = E^{(k)}$ and in $E^{(k)} = f_k(C)$ is different.* I will stick to the tradition of symbolising the quantity (size) of a magnitude q as $\{q\}$ and its quality as $[q]$, so that any magnitude can be represented as a unity of a quantity and quality.

Because, as shown above, the flipping of the sides of the equation $F = ma^{(2)}$ so that $a^{(2)} = F/m$ would hold cannot in fact be accomplished in classical mechanics without ending up in the triviality $a^{(2)} = a^{(2)}$, we immediately face the following problem. How is the concrete-to-concrete movement accomplished in classical mechanics if, on the one hand, that flipping cannot be performed in it, but on the other hand, some form of such a flipping has to be performed in that movement in classical mechanics; otherwise we cannot change the direction of our thought-movement from the concrete-to-abstract direction to that of the abstract-to-concrete direction. To emphasise this point let me utilise the conceptual differentiation between appearances and manifestations for the understanding of the sense and ambition of the movement N. Cartwright labelled as the concrete-to-concrete movement. Its ultimate sense and ambition is, first to reduce, in the course of the thought movement from the concrete to the abstract, the various effects-phenomena as appearances to their common cause, here expressed by means of $f_1(C) = E^{(k)}$, from which, then, these phenomena-effects ought to be derived as the manifestations of this cause and of its capacity (power), here expressed as $E^{(k)} = f_k(C)$, $E^{(k-1)} = f_{k-1}(C, C\mathrm{mod}_k), \ldots, E^{(k-j)} = f_{k-j}(C, C\mathrm{mod}_k, \ldots, C\mathrm{mod}_{k-j})$. So, it is readily seen that the change of the course in the movement concrete-to-abstract to that of abstract-to-concrete should be based on the flipping of $f_1(C) = E^{(k)}$ into $E^{(k)} = f_k(C)$, and what should be behind such a flipping should, of course, be the grasping of the very cause C – initially identified by means of $f_1(C) = E^{(k)}$ – *independently* from appearances (i.e., what the cause is in itself), and *prior* to the derivation of its manifestations $E^{(k)}, E^{(k-1)}, \ldots, E^{(k-j)}$. Let me, therefore, have a look first at Newton's concrete-to-concrete movement as he performs it in his *Principia*.

The *Principia* can, in my view, be understood as an attempt to unify the knowledge of the various physical phenomena according to a single basis, and then explain them on this basis (i.e., in terms used above, to transform them into manifestations of this ground). In fact, Newton, states this aim in the introduction to the first edition of the *Principia* : "the whole burden of philosophy seems to consist in this – from the phenomena of motion to investigate the forces of nature, and then from these forces to demonstrate the other phenomena [...]" (Newton, 1999, 382).

Let me, therefore, analyse Newton's approach to the concept of force. Force is for him the ground upon which physics should base its attempts to unify and explain the various phenomena of nature. Of crucial importance for us are definitions VI, VII and VIII (Newton, 1946, 4; Newton, 1999, 406–407):

i The absolute quantity of a centripetal force is the measure of the same, proportional to the efficacy of the cause that propagates it from the centre, through the spaces round about.

ii The accelerative quantity of a centripetal force is the measure of the same, proportional to the velocity which it generates in a given time.

iii The motive quantity of a centripetal force is the measure of the same, proportional to the motion which it generates in a given time.

Then, for the sake of brevity, he labels these magnitudes as the *absolute*, *accelerative*, and *motive* forces.

What unifies definitions VII and VIII is the fact that in both the quantity of the force is determined by the measure of the effects of this force, those effects being change of velocity in a given time and change of motion in a given time. These are, with respect to force, only its apparent (external) measures. Now what about the internal, inherent, absolute measure of the very force as it exists in itself, independently of phenomena like change of velocity or change of motion? In definition VI Newton explicitly mentions the absolute force and in the commentary to definition VIII he characterises it as a "cause [...] that does not yet appear" (Newton, 1946, 5; Newton, 1999, 407). But here he does not say what its internal measure is. In definitions VII and VIII something different from force – namely something external to it (change of velocity or change of motion in a given time) – is related to force as its own cause. But neither in definition VI nor elsewhere in the *Principia* does one find any reference to something different from force – something that would be internal with respect to force, i.e., related to it as its measure. On the contrary, he views the concept of absolute force as "purely mathematical, for I am not now considering the physical causes and sites of forces" (Newton, 1946, XVIII; Newton, 1999, 407).

What still can be calculated by means of the external measure of the force is the latter's quantitative determination (i.e., size). This is expressed by Newton in the claim that the purpose of the investigation of force (Newton, 1946, 550)

> is only to trace out the quantity and properties of this force from the phenomena [...], and to apply what we discover in some simple cases as principles, by which, in a mathematical way, we may estimate the effects thereof in more involved cases; [...] We said in a *mathematical* way, to avoid all questions about the nature or quality of this force, which we would not be understood to determine by any hypothesis [...]

One consequence of my change in Nowak's reconstruction of the structure of the idealizational law $L^{(k)}$ is that we have to modify also his reconstruction of the process of explanation by gradual concretization. Because this method proceeds, say, in classical mechanics from an idealizational law with the structure of 28 and not 7, the corrections, taking into account the causal impact of the modification conditions, *can be made only in respect to the effect-phenomenon* $E^{(k)}$ *as appearance.* The process of explanation by gradual concretization proceeding from a law of type $L^{(k)}$ with the structure of 28 leads to a law with the structure

$$(x)[Gx \& C\mathrm{mod}_k x \neq d_k \& \ldots \& C\mathrm{mod}_1 x = d_1 \\ \rightarrow E^{(k-1)}x = f_{k-1}(E^{(k)}x, C\mathrm{mod}_k x)]$$

finally to arrive at

$$(x)[Gx \& C\mathrm{mod}_k x \neq d_k \& \ldots \& C\mathrm{mod}_{k-j} x \neq d_k \\ \& C\mathrm{mod}_{(k-j-1)-1} x = d_{(k-j-1)-1} \\ \rightarrow E^{(k-j)}x = f_{k-j}(E^{(k)}x, C\mathrm{mod}_k x, \ldots, C\mathrm{mod}_{k-j} x)]$$

To prove my claims about the process of explanation based on a law with the structure of 28 hold, let me turn first again to Newton's *Principia* and then to contemporary classical mechanics.

How Newton deals with the relation of two bodies after he dealt with the notion of accelerative, motive and absolute force? Let me denote them as A and B. By choosing the former mass m_A as a unit it is possible to express the mass of the body B as a multiple or fraction of m_A. Body A is here in such a position that B can express its mass as equivalent or as a degree of equivalence with A. So A is here, with respect to B, in the position of an equivalent form. Body B is here in another position. What its mass is expressed with respect to (i.e., relatively to) A; it is in a relative form. This can hold also the other way round; one could take the mass of B and set it as a unit. So not only does A provide its body to express something belonging to B, but also B's body can be used to express something belonging to A. This something is, according to Newton, force f. In the former case force f_B tied to the body B uses the body A to manifest itself as a_A. In the latter case f_A tied to the body A uses the body B to manifest itself as a_B. In the latter case A is in the position of the relative form while body B is in the position of the equivalent form; in the former case it is the other way round. The basis of the fact that the mutual relation of A and B is that of the relation of an equivalent and relative form or vice *versa* is the existence of something different from acceleration and mass – it is force. It exists as a common ground in both bodies. What is common to both of them is one quality $[f]$ given in each of them in quantities $\{f_A\}$ and $\{f_B\}$, so that $f_A = f_B$ or $\{f_A\}[f] = \{f_B\}[f]$ holds. On both sides of this equation we have the same quality and the same quantity. This equation is, in my view, present in Newton's third law: "the mutual actions of two bodies upon each other are always equal, and directed to contrary parts" (Newton, 1946, 13; Newton, 1999, 417). This is readily seen from the commentary on the third law "If a body impinge upon another, and by its force change the motion of the other, that body also [...] will undergo an equal change, in its own motion, towards the contrary part" (Newton, 1946, 14; 1999, 417).

So this law in fact claims that "the mutual force-actions of two bodies are always equal and directed to contrary terms." What has to be emphasised is that because Newton in the third law and its commentary starts from the relation $f_A = f_B$, he can arrive at the claim that $dp_A/dt = dp_B/dt$ holds (i.e., that $m_A a_A = m_B a_B$ holds), and therefore finally also claim that "the changes of the velocities made towards contrary parts are inversely proportional to the bodies" (Newton, 1946, 14; Newton, 1999, 417). But this thought-movement bears in itself a fundamental problem. Newton, correctly (at least in my view) starts from the fact that bodies mutually interact, and through these interactions acquire certain phenomenal properties which are quantitatively determined. Then he tries to grasp the ground of these interactions by passing over to the concept of force. Finally, Newton wants to arrive at an explanation of these interactions by means of his three laws (axioms); i.e., to explain the phenomena as manifestations of force. In the first law the change of the state of a body *is reasoned by the concept of force* acting on it. In the second law the change of motion in time is also reasoned by the concept of force. Finally, in the third law the action and reaction of bodies in their mutual interaction is reasoned by the concept of force as well. But because Newton, as shown above, does not reason the quality of force, in his *Principia*, it remains a complete mystery why bodies mutually interact. Because he does not know what $[f]$ is, he has to claim in the second law, that "[t]he change of motion is proportional to the motive force impressed" (Newton, 1999, 417; Newton, 1946, 13). But because the motive force is defined in Definition VIII as proportional to the change of motion in time, Newton's movement from a phenomenon-effect as a *concrete appearance* given in this definition to the same phenomenon-effect as a *concrete manifestation* stated in the second law ends up in the trivial claim that the change of motion of a body in a certain time is proportional to the change of its motion in this time. For the same reason, namely, the absence of knowledge of what

$[f]$ is, Newton can neither derive the relation $\{f_A\}[f] = \{f_B\}[f]$ as it is given in the third law.

But even if mechanics cannot explain why force manifests itself by the phenomenon-effect of change of motion or of velocity, still knowing at least the quantitative determination (the size) of the acting force on the basis of the knowledge of the quantitative determinations of those effects-phenomena, it can on the basis of such a knowledge derive/explain the quantitative determination of *other* phenomena-effects. As an example of this let us take the case of the derivation of the law of free fall as well as the derivation of the concept of work, and, finally, the derivation of the size of the escape velocity of a body.

In the case of the derivation of the law of free fall the idea is that knowing the force acting on a body, say, on Earth, from its effect on this body – acceleration g – we can derive the path covered by that body under the influence of that force. If we suppose that on Earth the force acts on a falling body, which we initially regard as a mass-point, only in the direction of the y-coordinate, then we proceed from the following three differential equations: $F_x = md^2x/dt^2 = 0$; $F_y = md^2y/dt^2 = -mg$; $F_z = md^2z/dt^2 = 0$. By integrating them twice and by a suitable choice of the beginning of the coordinate system we derive the law of free fall. The corner-stone of the whole derivation of the *effect as manifestation* of F – the covered path – is the *prior* knowledge of the *effect as appearance* – g – of the force acting on the body in the y-coordinate. Because in Newton's second law we *cannot* characterise force F independently of its effect-appearance – acceleration g (on Earth) – but only by means of $F = mg$, we can neither substitute, say in the law of free fall or in any law derived from it by gradual concretization, the term g by the term F/m. Such a substitution would end up in the trivial substitution of g by g.

On the basis of Newton's second law, knowing the quantity of the acting force, we can determine the quantity of other effects it has on a body by means of the *path* on which it acts. The effect of the force $\mathbf{F}$ along a certain path is expressed as $\mathbf{F}d\mathbf{r}$, where $d\mathbf{r}$ is the elementary increase of the radius vector $\mathbf{r}$, giving the instant position of the moving body, viewed here again as a mass point. Integrating $\mathbf{F}d\mathbf{r}$ between two positions of that body, given as r_1 and r_2, yields the work A_{12} performed by F on that body between these two points. Because F has to be calculated by means of its effect, say, on earth, by $F = mg$, and supposing again that it acts only in the y-coordinate, we end up with $A_{12} = mg(y_1 - y_2)$. Again, in this equation and in any other derived from it by gradual concretization, we cannot substitute for g the expression F/m without ending up in the trivial substitution of g by g.

On the basis of $F = ma$ we can determine not only its effect along a certain path, but also – knowing the concrete character a path – determine the quantity of the force which generated this path. So, e.g., on the basis of $F = ma$ and Kepler's laws describing the characteristics of the path of a planet moving around its sun, we can determine the quantity of the force causing it, i.e., we can derive the law of gravity. This, in turn, enables us to determine the quantity of the effects of the gravity of this planet on a body, say, the quantity of circular velocity of this body around this planet in a certain distance from it. Knowing this quantity we can, then, determine the three cosmic velocities of this body: cosmic velocity one this body has to reach in order to move at the same circular speed as the circular velocity of the planet from which it was launched; cosmic velocity two it has to reach in order to escape the gravitational field of this planet; cosmic velocity three it has to reach in order to escape the gravitational field of that planet's sun.

The moral of these examples is, thus, the following. If the scientific law upon which the explanation of the effects rests has the character of 28, i.e., the starting equation has the structure $f_1(C) = E^{(k)}$, then the basis of explanation of the phenomena-effects as manifestation is *not* the very cause and its capacity (power), but in fact another phenomenon-effect which has the cognitive status of an appearance, by means of which we can determine at least the quantity (size) of the cause. For example, as shown above, if one takes the law of free fall as the basis of explanation by gradual concretization, then the *component permanently present in this explanation* is g, i.e., the effect-phenomenon as *appearance* of the acting force upon which we determine its quantity by means of $F = mg$.

After such a lengthy investigation into the character of the fundamental laws of mechanics I can, finally, deal with one argument of N. Cartwright from her *Lies*-period, with which I purportedly did not deal with above. In it she attacked the law of gravity and Coulomb's law, claiming that, without the *ceteris paribus* modifiers, they lie. The former one (restricting ourselves only to classical mechanics and electrostatic theory) is as follows:

$$(x)(y)(Ox\&Oy\&V_y = 0\&F_e xy\&qx = 0\&q'y = 0 \\ \rightarrow \mathbf{F}xy = Gmxm'y\mathbf{r}xy/r^3xy) \tag{29}$$

where O denotes an object with a non-zero mass; V denotes volume; F_e denotes external force acting on the physical system where x and y are placed; q and q' denote charge; m and m' denote mass. Coulomb's law can be written as

$$(x)(y)(O'x\&Oy\&V_x = 0\&V_y = 0\&F_exy\&mx = 0\&m'y = 0 \quad (30)$$
$$\rightarrow \mathbf{F}xy = qxq'y\mathbf{r}xy/r^3xy)$$

where O denotes objects with a non-zero charge. To express that two bodies interact both by electric and gravitational force we have to concretise 29 or 30. We obtain from the former

$$(x)(y)(Ox\&Oy\&V_x = 0\&V_y = 0\&F_exy\&qx \neq 0\&q'y \neq 0 \quad (31)$$
$$\rightarrow \mathbf{F}xy = Gmxm'y\mathbf{r}xy/r^3xy + qxq'y\mathbf{r}xy/r^3xy)$$

This means that the resultant force is given here by the vector addition of the gravitational and electric force. But, according to Cartwright (Cartwright, 1980a, 78–79; Cartwright, 1983, 59–60)

> [t]he vector addition story is, I admit, a nice one. But it is just a metaphor. *We* add forces (or the numbers that represent forces) when we do calculations. Nature does not 'add forces'. For the 'component' forces are not there, in any but a metaphorical sense, to be added; and the laws that say they are must also be given a metaphorical reading [...] These laws, I claim, do not satisfy the facticity requirement. They appear, on the face of it, to describe what bodies do: in one case the two bodies produce a force of size Gmm'/r^2, in the other a force of size qq'/r^2. For the force of size Gmm'/r^2 and the force of size qq'/r^2 are not real, occurrent forces. In interaction, a single force occurs – the force we call the 'resultant' – and this force is neither the force due to gravity nor to electric force. On the vector addition story, the gravitational and the electric force are both produced, yet neither exists.

Now what strikes us here immediately is her claim about the production of forces. But as shown above, the law of gravitation is derived in classical mechanics from the knowledge of the path of a planet around its sun and the law $F = ma$; *in classical mechanics there is not even a trace of a knowledge about the production or origin of the force of gravity.* Similarly, in the case of the electric force by using a torsion balance we determine the force acting between two small charged balls by first determining the torque *effect* of that force on the fibre attached to the arm connecting the small balls. Like in classical mechanics, *there is no way in electrostatic theory to determine the size of the electric force independently of its effects*, i.e., *in this theory there is no understanding of the very process of production of electric force.*

What also has to be emphasised here is that even if N. Cartwright correctly, in my view at least, claims that we have in the case of the mutual interaction of charged bodies with a non-zero mass just *one* force, still we lack in physics a unified theory of the electro-gravitational force.

On the one hand, electrostatic theory became, via the theory of electromagnetism, part of the quantum field theory while Newton's law of gravitation became part of the general theory of relativity. But on the other hand, we still lack, to best of my knowledge, a unified theory which would give us at least a receipt for the calculation of the quantity (size) of *one* electro-gravitational force. Therefore also, in the process of gradual concretization from 30 to 31 given above, the additional component appearing in the consequent of 31 is *not* derived on the basis of knowledge from *classical mechanics* about the impact of acting modification conditions – where this acting is expressed as $q \neq 0$ and $q' \neq 0$ – but is simply put in on the basis of the knowledge of Coulomb's law from a completely another theory, namely, electrostatic theory.

References

Cartwright, N. (1979). Causal laws and effective strategies. *Noûs*, 13:419–437.

Cartwright, N. (1980a). Do the laws of physics state the facts? *Pacific Philosophical Quarterly*, 61:75–84.

Cartwright, N. (1980b). The truth doesn't explain much. *American Philosophical Quarterly*, 17:159–163.

Cartwright, N. (1982). When explanation leads to inference. *Philosophical Topics*, 13:111–121.

Cartwright, N. (1983). *How the Law of Physics Lie*. Clarendon, Oxford.

Cartwright, N. (1985). Causation in physics: Causal processes and mathematical derivations. In of Science Association, P., editor, *PSA Proceedings 1984*, volume 2, pages 391–404, East Lansing.

Cartwright, N. (1986). Fitting fact to equations. In Grandy, R. E. and Warner, R., editors, *Philosophical Grounds of Rationality*, pages 441–453, Oxford. Clarendon.

Cartwright, N. (1987). Philosophical problems of quantum theory. In Krüger, L., Daston, L. and Heidelberger, M., editors, *The Probabilistic Revolution*, Cambridge, London. MIT Press.

Cartwright, N. (1988a). Regular association and singular causes. In Skyrms, B. and Harper, W., editors, *Causation, Chance, and Credence*, pages 79–97, Dordrecht. Kluwer.

Cartwright, N. (1988b). Ursachen und mathematische Physik. In Muschik, W. and Scheibe, E., editors, *Philosophie, Physik, Wissenschaftsgeschichte*, pages 12–33, Berlin. Technische Universität Berlin.

Cartwright, N. (1989a). Capacities and abstractions. In *Minnesota Studies in the Philosophy of Science*, volume 13, pages 349–356. University of Minnesota Press, Minnesota.

Cartwright, N. (1989b). *Nature's Capacities and their Measurement.* Clarendon, Oxford.

Cartwright, N. (1990). The Born–Einstein debate. *Synthese*, 81:271–282.

Cartwright, N. (1991a). Can wholism reconcile the inaccuracy of theory with the accuracy of prediction? *Synthese*, 89:3–13. (quoted from the manuscript).

Cartwright, N. (1991b). Fables and models. *Proceedings of the Aristotelian Society*, XLV, Suppl.:55–68.

Cartwright, N. (1992). Aristotelian natures and the modern experimental method. In Earman, J., editor, *Inference, Explanation and Other Philosophical Frustrations*, pages 44–71, Berkeley. University of California Press.

Cartwright, N. (1993a). How we relate theory to observation. In Horwich, P., editor, *World Changes*, pages 259–273, Cambridge (MA). MIT Press.

Cartwright, N. (1993b). In defense of 'this worldly' causality. *Philosophy and Phenomenological Research*, 53:423–429.

Cartwright, N. (1993c). Is natural science 'natural' enough? *Synthese*, 94:291–301.

Cartwright, N. (1994). Fundamentalism vs. patchworks of laws. *Proceedings of the Aristotelian Society*, pages 279–292.

Cartwright, N. (1995a). Ceteris paribus laws and socio-economic machines. *The Monist*, 78:276–294.

Cartwright, N. (1995b). False idealisations. *Philosophical Studies*, 74: 339–352.

Cartwright, N. (1995c). Precise on nature's capacities and their measurement. *Philosophy and Phenomenological Research*, 55:153–156.

Cartwright, N. (1995d). Reply to Eells, Humphreys and Morrison on natures capacities and their measurement. *Philosophy and Phenomenological Research*, 55:177–187.

Cartwright, N. (1996). What is a causal structure. In Vaughn, R., McKim, R.,and Turner, S. P., editors, *Causality in Crisis?*, pages 343–357, Notre Dame. University of Notre Dame Press.

Cartwright, N. (1997a). Models: The blueprints for laws. *Philosophy of Science*, 67 (Suppl.):292–303.

Cartwright, N. (1997b). Where do laws of nature come from. *Dialectica*, 51:65–78.

Cartwright, N. (1998). How theories relate. *Philosophia Naturalis*, 35:23–34.

Cartwright, N. (2000). *The Dappled World.* Cambridge University Press, Cambridge.

Cartwright, N. and Mendel, H. (1984). What makes physics' objects abstract? In et al., J. T. C., editor, *Science and Reality*, pages 134–152, Notre Dame. University of Notre Dame Press.

Cartwright, N. and Nordby, J. (1983). How approximations take us away from theory and towards the truth. *Pacific Philosophical Quarterly*, 64:273–280.

Friedman, M. (1974). Explanation and scientific understanding. *Journal of Philosophy*, 71:5–19.

Hanzel, I. (1999). *The Concept of Scientific Law in the Philosophy of Science and Epistemology.* Kluwer, Dordrecht.

Hempel, C. G. (1965). *Aspects of Scientific Explanation.* The Free Press, New York.

Hempel, C. G. and Oppenheim, P. (1948). Studies in the logic of explanation. *Philosophy of Science*, 15:135–175. (Reprinted in Hempel, 1965: 245–290).

Messiah, A. (1961). *Quantum Mechanics*, volume II. North Holland, Amsterdam.

Newton, I. (1946). *Sir Isaac Newton's Mathematical Principles of Natural Philosophy and his System of the World.* University of California Press, Berkeley. (Translation A. Motte, revised by F. Cajori).

Newton, I. (1999). *The Principia.* University of California Press, Berkeley. (Translation by I. B. Cohen and Anne Whitman).

Nowak, I. (1974). The conception of dialectical correspondence. In Kmita, J., editor, *The Methodological Implications of Epistemology*, pages 252–256, Warszawa. PWN. (In Polish language).

Nowak, L. (1972). Laws of science, theory, measurement. *Philosophy of Science*, 39:192–201.

Nowak, L. (1980). *The Structure of Idealisation.* Reidel, Dordrecht.

Salmon, W. (1990). *Four Decades of Scientific Explanation.* University of Minnesota Press, Minneapolis.

THE EXPLANATORY VIRTUES OF PROBABILISTIC CAUSAL LAWS

Henrik Hållsten

1.

If there are probabilistic causal laws, what explanatory virtues do these laws have? Or to be more exact what explanatory virtues does a probabilistic cause have? It is often argued that if the world is indeterministic then there exist probabilistic causal relations, and these causal relations should be enough for an explanation. Writers such as Salmon, Railton and Humphreys argue that the fact that a cause is probabilistic should not matter for whether or not it suffices as an explanation of the effect. In the following I will argue against this position. Closely tied to this issue is the question of explanatory deductivism. If one wants to admit explanatory validity to probabilistic causes one has to give up the requirement that an ideal explanation consists of a deductive argument. Thus, if you believe that a probabilistic cause explains its effect, then you are a non-deductivist. This paper is part of an overall attempt to defend explanatory deductivism, but its concern is not to spell out deductivism but to elaborate on certain unwanted consequences of granting probabilistic causes explanatory validity.

Before we go on we need to briefly say something about *explanation* and about *probabilistic cause.* We will assume an objective model of explanation: that is, although different parts of an ideal explanation is relevant in different contexts a necessary condition for relevance in *any* context is objective relevance. The criteria for objective relevance are *not* epistemically relativized. Thus, if something is an explanation in a particular state of knowledge, it will still be objectively relevant in any conservative extension[1] of that state of knowledge. This position will not be defended here, but simply taken for granted.

[1]A conservative extension is taken to mean that only additions and no revisions has changed the epistemic state.

J. Faye, P. Needham, U. Scheffler and M. Urchs, (eds.), Nature's Principles, 137–150.

Not only is a probabilistic cause not sufficient for its effect, its is neither necessary. Although it is its lack of being sufficient that makes it probabilistic its lack of being necessary comes as part of the package deal. The fact that probabilistic causes are not necessary will also be referred to as the fact that they violate the *sine qua non* requirement. Neither this will be argued for here.[2]

2.

Take the example of operation Smoky in 1957, in which a number of American soldiers witnessed the detonation of an atomic bomb at close range (Salmon, 1988, pp. 160–161). There were 2235 participants, eight of whom developed leukaemia. This has been considered as an example of where the probabilistic cause explains its effect, in spite the fact that the cause was not sufficient and probably not necessary for the effect. Salmon argues that, for each one of the leukaemia-stricken soldiers the leukaemia can be explained with the soldier's participation in operation Smoky, in spite the fact that the cause was not sufficient and probably not necessary.

Assuming that the underlying mechanisms are truly indeterministic, we will in the following examine this claim. First of all, we have two alternatives. First, exposure to ionizing radiation might be the only way to contract leukaemia. Exposure is then a necessary condition and, for the sake of argument, let us assume that the only time the soldiers could have been exposed was while witnessing the detonation. Here it is tempting to admit that we do explain the sickness by pointing to the detonation of the bomb.

In such an event where we know that a specific probabilistic cause is a necessary condition for its effect, the arguments in this paper will be impotent. However, such examples are not easy to come by. We will therefore leave this alternative with noting that arguments against the position of admitting necessary probabilistic causes as explanations can be found in (Hållsten, 2001).

Having said this let us assume, a little more realistically, that there are other ways of contracting the disease; for example by genetic dispositions, exposure to chemicals or even viruses.[3] There would probably

[2]See for example §§7 and 9 in Humphreys, 1989 for arguments to the conclusion that probabilistic causality violates the *sine qua non* requirement. In fact, I know no model of probabilistic causality that contains the requirement of *sine qua non*.

[3]In *Scientific American, Sept. 1996* the following "Risk factors" are listed: "Certain genetic abnormalities, including Down's syndrome, Bloom's syndrome and ataxia-telangiectasia; excessive exposure to ionizing radiation and some chemicals, such as benzene, found in lead-free

be synergetic effects, but I think it is safe to say that there would be at least two different, although complex, ways of getting leukaemia. It would then be a very odd position not to consider the way the disease was contracted as a matter of fact, though it might be beyond our means of discovery. If we have two candidates, then either one or the other is responsible. And if we do not know which one it is, we should restrain from calling what we have a genuine explanation. If we list both candidates we would certainly list one too many and if we list just one we run the risk of listing the wrong one.

3.

In (Hållsten, 1999) I tried to argue the point above. Learning from my critics I will in the following try to elaborate on the argument as well as adding a new twist to it.

Let us assume that the mechanisms behind leukaemia can be so far simplified as to leave us with two alternative causal chains leading to the same effect, namely the occurrence of the disease in a specific person. One chain stems from genetic disposition in conjunction with exposure to radiation and the other from viruses in conjunction with exposure to chemicals. Let us schematize this as follows:

$$\begin{array}{ll} \left(\begin{array}{l}\text{exposure to ionizing radiation}\\ \text{genetic dispositions}\end{array}\right) & e_{1.1}\stackrel{\smile}{\to}e_{1.2}\stackrel{\smile}{\to}e_{1.3}\stackrel{\smile}{\to}e_{1.4}\stackrel{\smile}{\to} \\ \left(\begin{array}{l}\text{exposure to chemicals}\\ \text{exposure to viruses}\end{array}\right) & e_{2.1}\stackrel{\smile}{\to}e_{2.2}\stackrel{\smile}{\to}e_{2.3}\stackrel{\smile}{\to}e_{2.4}\stackrel{\smile}{\to}e_{2.5}\stackrel{\smile}{\to} \end{array} \quad l \qquad (1)$$

where l is a sentence of the form "Soldier so and so contracted leukaemia such and such there and then". Similarly, the sentences of the form $e_{n.m}$ all describe conditions that must be fulfilled for the chain to be unbroken. The probabilistic causal relation is written "$\stackrel{\smile}{\to}$", and the time being no more assumptions are made, apart from that if 1 constitutes a full description of the causal structure, then $e_{n.m}$ is a necessary, but not sufficient, condition for $e_{n.m+1}$, and $e_{1.4}$ and $e_{2.5}$ are neither sufficient not necessary conditions for l. This schema is admittedly very much simplified, but I cannot see how any complication making it more realistic would enhance the non-deductivist position that we are to discuss. More complications would probably just make it more likely that we are, in fact, dealing with an epistemically relativized explanation. If all we

gasoline; exposure to the virus HTLV-I." (Rennie, 1996, p. 97). I interpret "Risk factors" not as being non-causal risk factors, but as explicitly causal factors.

know is that the starting-nodes as well as the end-node are true, we do not know whether both or just one causal chain actually was responsible for bringing about leukaemia. If $e_{1.3}$ were false – something that is entirely consistent with $e_{1.2}$ being true, since the connections are indeterministic – that chain would be terminated, leaving the chain starting with $e_{2.1}$ as the only possible cause. Before we discuss this further and elaborate on it, let us now consider how to calculate the probabilities in a chain of events.

If we have a chain of events, $e_1 \breve{\rightarrow} e_2 \breve{\rightarrow} \ldots \breve{\rightarrow} e_k$, where every prior event is a necessary condition for the events that follow, the probability of the end, e_k, given the start of the chain, e_1, is a function of the conditional probabilities of all intermediate events. Consider three events e_1, e_2 and e_3, where e_1 is a necessary condition for e_2 and e_2 is a necessary condition for e_3 (e_1 is then a necessary condition for e_3). We want to calculate the conditional probability of e_3 given e_1 on the basis of the probabilities e_2 given e_1 and e_3 given e_2. We must first recognise that if an event is a necessary condition for β then $p(\alpha \wedge \beta) = p(\beta)$. This becomes obvious when we consider that the only time β is true is when both α and β are true. Second, we use the definition of conditional probability:

$$p(\alpha \mid \beta) = \frac{p(\alpha \wedge \beta)}{p(\beta)} \tag{2}$$

Thus,

$$p(e_3 \mid e_1) = \frac{p(e_1 \wedge e_3)}{p(e_1)} = \frac{p(e_1 \wedge e_2 \wedge e_3)}{p(e_1)}$$

since e_2 is necessary for e_3. We then use 2 again and obtain:

$$p(e_1 \wedge e_2 \wedge e_3) = p(e_1 \wedge e_2) \cdot p(e_3 \mid e_1 \wedge e_2)$$

And since e_1 is necessary for e_2,

$$p(e_3 \mid e_1 \wedge e_2) = \frac{p(e_1 \wedge e_2 \wedge e_3)}{p(e_1 \wedge e_2)} = \frac{p(e_2 \wedge e_3)}{p(e_2)} = p(e_3 \mid e_2)$$

Thus,

$$p(e_3 \mid e_1) = \frac{1}{p(e_1)} \cdot p(e_1 \wedge e_2) \cdot p(e_3 \mid e_2) = p(e_2 \mid e_1) \cdot p(e_3 \mid e_2)$$

Were we to expand the chain and make e_3 a necessary condition for e_4, this argument can be expanded to obtain $p(e_4 \mid e_1) = p(e_2 \mid e_1) \cdot p(e_3 \mid e_2) \cdot p(e_4 \mid e_3)$. Thus, if e_i is a necessary condition for e_{i+1}, then:

$$p(e_k \mid e_1) = p(e_k \mid e_{k-1}) \cdot p(e_{k-1} \mid e_{k-2}) \cdot \ldots \cdot p(e_2 \mid e_1) \qquad (3)$$

Thus, if we only know that $e_{1.1}$ and $e_{2.1}$ are true as well as the values of $p(e_{1.2} \mid e_{1.1}), \ldots, p(l \mid e_{1.4})$, and $p(e_{2.2} \mid e_{2.1}), \ldots, p(l \mid e_{2.5})$, the probability of leukaemia is a function of all these independent probabilities concerning the two chains in 1. This probability of getting leukaemia is obviously relative to a specific state of knowledge, namely a state consisting of $e_{1.1}$ and $e_{2.1}$ but that says nothing conclusive about whether $e_{1.2-4}$ or $e_{2.2-5}$ are true. Were we to know anything about these other links in the two chains, our estimation of the probability would change. Obviously, we do not have to consider $p(e_{1.2} \mid e_{1.1})$ when calculating $p(l \mid e_{1.4} \wedge e_{2.5})$, so if we know that both chains are unbroken until $e_{1.4}$ and $e_{2.5}$, the important probabilities are $p(l \mid e_{1.4})$ and $p(l \mid e_{2.5})$. (We will return to the calculation of $p(l \mid e_{1.4} \wedge e_{2.5})$ on the basis of $p(l \mid e_{1.4})$ and $p(l \mid e_{2.5})$ later in this section.)

It is thus clear that any statistical explanation of l that is founded on the probabilities $p(l \mid e_{1.1})$ and $p(l \mid e_{2.1})$ would be epistemically relativized since more knowledge might $e_{2.2}$ show to be false making $e_{2.1}$ objectively explanatory irrelevant. Had $e_{2.1}$ failed to bring about $e_{2.2}$, $e_{2.1}$ would have been irrelevant to the explanation since it actually had not helped bring l about. And even if we, like Salmon, where to build our explanation on the causal structure, listing $e_{1.1}$ and $e_{2.1}$ would hardly be anything more than a sketch or a part of an inductive process leading to a genuine explanation.

A more complicated case arises where all nodes in the chains are true, *and known to be so.* It is only when we know that both chains are unbroken – in the sense that all nodes are true – that further knowledge can be gained without the possibility of showing one of the chains to be irrelevant. It would be unrealistic to claim that this is the epistemic situation concerning operation Smoky. In fact, this seems to be so for all examples from applied science. But, for the sake of argument, let us suppose that we know that none of the chains is broken. Assume in the following that we know both $e_{1.4}$ and $e_{2.5}$ in 1 to be true.

Since both links to l are indeterministic, had we not yet known about the leukaemia we would not have been able to predict it with 100% certainty. The question is, if l is true, do $e_{1.4}$ and $e_{2.5}$ constitute the explanation l? My answer is still no.

My claim in this case is that there would still be three ways for the effect to come about: because of one or the other of the alternative chains, or of both together. When the effect comes about, there will be no way of deciding which actual causal path was effective. The reason for this is that if there exists a method to decide in retrospect which of the three ways that was effective, this method would have to build on information contained in the effect. But if such information exists, we would be justified in individuating three possible effects instead of one – $e_{1.4}$ would have brought about l_1, $e_{2.5}$ l_2, and had they both been effective $l_{1\&2}$ would have been the case. Since, according to the premises, there was only one probable effect, no such method exists. Thus, for the situation where both $e_{1.1}$, $e_{2.1}$ and l are true there are still three ways, or causal paths, in which $e_{1.1}$ and $e_{2.1}$ might have caused l, but we cannot know which path is taken.

Since we cannot know which one of the presumed causes really was effective, the listing of them both should not be regarded as a genuine explanation in spite of the fact that this is the best our theories can do. The problem is that probabilistic causal relations violate, as noted, the *sine qua non* requirement. Literally, this requirement states that if the cause would not have been, neither would the effect, or more precisely, if *this* cause would not have been, neither would *this* effect. — With another cause we would, under the same circumstances, have got another effect. But since probabilistic causality violates this requirement, had not one of the causes, $e_{1.1}$ and $e_{2.1}$, been the case, the effect, l, could still have been brought about under the same circumstances on account of the other cause.

If the non-deductivist insists that it is the *conjunction* $e_{1.1}$ and $e_{2.1}$ that constitutes the explanation, then he must show that they did in fact work together. To defend the explanatory completeness of these two *sine qua non* violating probabilistic mechanisms, one would have to argue against the three-way possibility of getting the effect. It must be shown to be impossible for the two alternatives to do anything other than work together. And this must be done in spite of the fact that one of them could have been enough, as well as the fact that both of them are not sufficiently effective at all times. We know that $e_{1.4}$ could have caused l on its own and that it could have failed to do so. And we know that the same holds for $e_{2.5}$. What the non-deductivist must show then, is that when l does get caused and both $e_{1.4}$ and $e_{2.5}$ are true, then they are both effective.

To show this it must be demonstrated that the effectiveness of the two alternative causal chains is correlated in such a way that one is effective if and only if the other is effective. If this cannot be achieved, we should adhere to deductivism and refrain from admitting probabilistic causes as explaining their effects.

The above is merely an elaboration on the argument as it was presented in (Hållsten, 1999). However, a conversation over a beer and goulash with Lars Bergström convinced me that I had to give it a new twist. When I was reluctant to accept that the two candidates *must* cooperate, he responded that perhaps we should for the sake of simplicity not consider them to be two causes but one cause. As one cause, it is obvious that it works as a whole. One way of responding to this is to try to elaborate on the picture 1. Instead of having $e_{1.4}$ and $e_{2.5}$ we would then have a conglomerated event $e_{1\&2}$ causing l. But if this gives us the following scheme;

$$\begin{array}{r} e_{1.1} \stackrel{\smile}{\rightarrow} e_{1.2} \stackrel{\smile}{\rightarrow} e_{1.3} \stackrel{\smile}{\rightarrow} \\ e_{2.1} \stackrel{\smile}{\rightarrow} e_{2.2} \stackrel{\smile}{\rightarrow} e_{2.3} \stackrel{\smile}{\rightarrow} e_{2.4} \stackrel{\smile}{\rightarrow} \end{array} \quad e_{1\&2} \stackrel{\smile}{\rightarrow} l \tag{4}$$

then surely nothing is gained. In 4 the same question can be asked concerning $e_{1.3}$, $e_{2.4}$ and $e_{1\&2}$ as has been pursued concerning $e_{1.4}$, $e_{2.5}$ and l in 1. Instead, in order to understand this objection, it is the key notion probabilistic causality, or probabilistic mechanism, that must be elaborated. The following proposal tries to do this without relying on any of the specific models of probabilistic causality as they have been presented.[4]

When considering the relation denoted by '$\stackrel{\smile}{\rightarrow}$' above, I considered it to hold on account of what will here be called a *chancy mechanism*. The effectiveness of the mechanism, or causal relation, is chancy, i. e. subject to chance. This chancy mechanism sometimes fails to bring its effect about, but sometimes it succeeds. It is a genuinely chancy mechanism in the sense that we cannot know whether it will fail or succeed. If we know that the cause is present, and hence the mechanism is working, we will still not know if this will really bring about the effect, and there will be nothing more to know which could enhance our prediction. When I questioned the explanatory validity of probabilistic mechanisms above and in (Hållsten, 1999), I was clearly basing my argument on the chancy mechanism-view of probabilistic causality. The two alternative causes are then seen as two guns with which we simultaneously play Russian roulette. Using both, the subjective probability increases but their respective chanciness constitute two separate issues. Thus, the argument holds if we view probabilistic mechanisms as chancy mechanisms.

The other way of viewing this matter is as a *chance mechanism*, where the mechanism always succeeds in producing a chance (or propensity), that is an objective feature of the world (what kind I dare not say). When this chance obtains, there will be a likelihood that the effect will take place. If we know the strength of the chance, this likelihood will be the strength of our rational expectation, i.e. the subjective probability

[4]See for example (Suppes, 1970), (Salmon, 1980) or (Salmon, 1984), (Humphreys, 1989), (Eells, 1991) or (Mellor, 1995).

of the effect. Our knowledge of chances thus determines our subjective probabilities. As Brian Skyrms puts it:

> If, for each chance hypothesis, we have the degree-of-belief probability of the observed outcome sequence conditional on that chance hypothesis, then we can use Bayes' theorem to update our degree of belief about chance. The natural thing to do – and what everyone does do – is to take the (degree of belief) probability of an outcome sequence conditional on a chance hypothesis to be equal to what the chance hypothesis says it is. (Skyrms, 1992, p. 377)

Given that the concept of event is so nebulous, chances could easily be understood as a peculiar kind of event. Hence, if the strength of a chance of an event e is a, then the probability of that event occurring due to this chance is calculated according to

$$p(e \mid \text{chance-of- } e = a) = a \tag{5}$$

This is in accordance with Paul Humphreys' basic ideas concerning probabilistic causality:

> This appeal to an increase in the probability of the effect is the starting point of all theories of probabilistic causality. Noting that what is produced is not directly the effect, or changes in it, but a change in its probability, I propose to take this idea seriously – in fact, literally – and see what ensues. That is, let us allow provisionally that the chance of a phenomenon is measurable, and moreover something which can be increased or diminished in value by changes in certain properties. Then using our basic principle, we have as a preliminary working hypothesis that a factor X is a (probabilistic) contributing cause of a factor Y if and only if X's existence contributes to the chance of Y's existence. (Humphreys, 1989, pp. 14–15)

I therefore take Bergström's objection to be directed against the chancy mechanism view of probabilistic causality and to support the chance mechanism view. Thus, the two alternative causes, $e_{1.4}$ and $e_{2.5}$, both contribute to the combined chance of the effect. They are both always effective in bringing about this addition, and should hence both be present in the explanation should the effect be actual.[5]

The case is closed, and the explanatory validity of probabilistic causality is saved in spite of its non-adherence to the requirement of *sine qua non*, you might say. But is this really true? Can we not ask a similar question concerning chances as we did concerning (chancy) mechanisms? Let us do that, and add a further twist to the argument.

[5]Observe that the requirement of *sine qua non* does not hold under this interpretation of probabilistic causality. In the absence of one of the co-operating causes, the other one could have brought the effect about.

A chance is thus an objective feature of the world, a sort of (quasi-?) event brought about by a chance mechanism, in virtue of which the mechanisms tend to bring about the effect. If c_1 and c_2 are two independent chance mechanisms[6] for e, then c_1 gives rise to a chance of e, as does c_2. Let us call these the chance-of-e-due-to-c_1 and the chance-of-e-due-to-c_2. It is obvious that they both contribute to the *subjective* probability of e. This subjective probability can be obtained by first using 5 to calculate $p(e \mid c_1)$ and $p(e \mid c_2)$. The next step is to calculate the probability of e due to both independent causes c_1 and c_2: $p(e \mid c_1 \wedge c_2)$. Since, $p(e \mid c_1 \wedge c_2) + p(\neg e \mid c_1 \wedge c_2) = 1$,

$$p(e \mid c_1 \wedge c_2) = 1 - p(\neg e \mid c_1 \wedge c_2)$$

And, since c_1 and c_2 are independent, the probability of $\neg e$ given $c_1 \wedge c_2$ equals the probability of $\neg e$ given c_1 multiplied with the probability of $\neg e$ given c_2. Thus,

$$p(e \mid c_1 \wedge c_2) = 1 - p(\neg e \mid c_1) \cdot p(\neg e \mid c_2) \tag{6}$$

And since $p(\neg e \mid c) = 1 - p(e \mid c)$,

$$\begin{aligned} p(e \mid c_1 \wedge c_2) &= 1 - (1 - p(e \mid c_1)) \cdot (1 - p(e \mid c_2)) \\ &= 1 - 1 + p(e \mid c_1) + p(e \mid c_2) - p(e \mid c_1) \cdot p(e \mid c_2) \end{aligned}$$

which reduces to the more familiar formula:

$$p(e \mid c_1 \wedge c_2) = p(e \mid c_1) + p(e \mid c_2) - p(e \mid c_1) \cdot p(e \mid c_2) \tag{7}$$

(Råde and Westergren, 1988, p. 295).

This is straightforward. But if the view of chance mechanisms is to refute the argument here and in (Hållsten, 1999) any better than chancy mechanisms do, it is necessary that the *chances* add. And if they add up in a similar way as do probabilities we must ask *why*? When we have two independent causes, the subjective probability of the non-occurrence is the probability of neither of the two causes bringing about the effect – i.e. not of one alone, nor of the other alone, nor of both together. And, as we have seen above in 6, the subjective probability of the occurrence is 1 minus this. *So the way we add subjective probabilities mirrors the view that there are three ways of causing the effect.*

If chances are objective features of the world, what argument can be presented against simply adding them according to

[6] Let us say that a mechanism is individuated by its cause node.

$$\text{chance-of-}e\text{-due-to-}c_1\text{-and-}c_2 = \text{chance-of-}e\text{-due-to-}c_1 + \text{chance-of-}e\text{-due-to-}c_2 \quad (8)$$

instead of mimicking subjective probability:

$$\text{chance-of-}e\text{-due-to-}c_1\text{-and-}c_2 = \text{chance-of-}e\text{-due-to-}c_1 + \text{chance-of-}e\text{-due-to-}c_2 - \text{chance-of-}e\text{-due-to-}c_1 \cdot \text{chance-of-}e\text{-due-to-}c_2? \quad (9)$$

Admittedly, we would have to modify 5 if we chose 8[7], but could not two 50% chances together determine the effect? The only reason to opt for 9 seems to be that *what we really add is (subjective) probabilities, not chances.*

To calculate the subjective probability of the effect according to 7, is to admit that there are four possible alternatives when both c_1 and c_2 are true. Either (i) e is true, due to both the chance-of-e-due-to-c_1 and the chance-of-e-due-to-c_2, (ii) e is true, due only to the chance-of-e-due-to-c_1, (iii) e is true, due only to the chance-of-e-due-to-c_2, or (iv) e is false, since the effect fails in spite of the chance-of-e-due-to-c_1 and the chance-of-e-due-to-c_2. If chances do add, why should they do so in accordance to this?

Let us summarize. If we were to adopt the chancy mechanism view, there would clearly be a matter of fact as to which one really caused the effect, and if we were to adopt the chance mechanism view, unless it is shown that objective chances add, we would reach the same conclusion.

The non-deductivist's last resort is to maintain that chances do add in this most peculiar way. Just like forces, they always add, but they do so in a manner totally different from forces. When forces add they bring about a modified event. (If we all push, the wagon moves faster, or the brakes become hotter, than if just one of us had applied force.) Chances – or propensities – do not bring about anything new when added, they just bring the old effect about with a higher probability. And, as noted above, they add up to that combined chance just like subjective probabilities add. Perhaps that is just what chances – or propensities – do. But we have no way of deciding if they really do this or not; thus this hypothesis would only increase our metaphysical burden without enhancing our scientific theory. It would be less metaphysically extravagant to claim that chances do not add, although subjective probability does – i. e. the

[7] We could simply add the constraint that 5 holds if $a \leq 1$, and that if $a > 1$ then $p = 1$

degrees of our rational beliefs add, and this is done in a non-controversial way.

Given this less extravagant view of chances, the deductivist has something that can be explained by the D-S explanation as well as an argument against non-deductivism.

4.

Still there are rejoinders to be found. In (Humphreys, 1989), §14, a similar case as above is discussed. James Woodward, in personal conversation with Humphreys, raised an argument basically equivalent to the argument above.

> Consider, he [Woodward] suggests, the example that I [Humphreys] used in §9 of the mouse exposed to two carcinogenic chemicals. Given that the mouse develops a stomach tumor, and that each chemical increases the chance of contracting such a tumor, he suggests that this information alone does not discriminate between the following three possibilities: (1) the tumor is entirely caused by the first chemical, (2) the tumor is entirely caused by the second tumor, (3) both chemicals contributed causally to the tumor. (Humphreys, 1989, p. 36)

Humphreys' reply consists in first noting that according to his theory of causality, the propensity to which a causal factor contributes does not in itself cause the effect.

> [O]nce the contribution of B [a causal factor] has been cited, there is nothing more to say that is genuinely causal – it is simply that the higher the value of the propensity, the more likely A [the effect] is to occur. (Humphreys, 1989, p. 36)

The propensity of A is thus what is increased or decreased by the causal factors. There is – according to Humphreys' theory – no propensity-of-A-due-to-B. Concerning the mouse tumor with two causes, Humphreys states:

> The correct response to the example is this: probability is construed here in terms of single case chances [...]. In the situation at hand, the situation with both carcinogens present *is* different from the situation with only one – the chance is higher than with either alone because both chemicals have contributed to the value of that chance. And that is all there is. (Humphreys, 1989, p. 36)

And:

> The second chemical is not irrelevant on this (or on any other occasion) for it contributes to the chance on this occasion, as does the first chemical, and after they have done this, *nothing else causally happens.* (Humphreys, 1989, p. 37)

First, it is not obvious that the considerations in the beginning, concerning situations where we do not have full knowledge, do not also apply here. If the explanation is not to be epistemically relativized – and hence not genuine – we must *know* that both causal chains are unbroken from the presence of the two chemicals to the last step before the tumor. (See 1 – we must know that the corresponding $e_{1.4}$ and $e_{2.5}$ both are true.) If one of the chains were terminated before the tumor appeared, then that chemical would not have anything to do with the tumor. A model of explanation that is true to the idea of objective relevance cannot count both carcinogens as explanatory simply because we do not yet now that one of the chains are terminated. If Humphreys' model cannot discriminate this situation, so much worse for the model.

We get a more complicated situation if we assume that everything that is relevant to the effect is known. Let us therefore, once again, consider the case where we know that both chains are unbroken until the last events before the tumor. Here we can see that there is a distinctive feature in Humphreys' model as opposed to the two ways in which I have pictured probabilistic causality in chancy mechanisms and chance mechanisms. For Humphreys, there is no mediating feature between the two causal factors and the propensity of the effect. Causal factor x does not cause a propensity-due-to-causal-factor-x, which is then added to the other propensities. When discussing how changing the situation from $\neg B$ to B causes (probabilistically) A, Humphreys states:

> This [...] can be interpreted within our approach in terms of the effect that the chance from $\neg B$ to B has on the propensity of the system to produce A. The change deterministically produces an increase in the propensity from its base level of $p(A \mid \neg B)$ to a new value $p(A \mid B)$ by virtue of the contribution $R(A, B)$. (Humphreys, 1989, pp. 34–35)

The contribution $R(A, B)$ is called "the relevance difference" (p. 34). Thus, Humphreys circumvents all talk about propensities as being relative to what causes them. Changing $\neg B$ to B gives a contribution the propensity of the system to become A, and if changing $\neg C$ to C also has this effect it just pours a bit more into the bucket of propensity. If A then does become true, it does so on account of both contributions.

It thus might seem like Humphreys (if we grant him that we know that none of the chains are broken until the last event) can get past the argument as presented with chance-mechanisms. But being stubborn, we might ask the same question about propensity contributions as we did about chances, and that is: *Why do they add like probabilities, in accordance with 7?* As we have seen, the rationale behind 7 is the three-way possibility of causing the effect. If Humphreys wants to dismiss this three-way possibility, why do his causality contributions add in accordance with it? Just as for the chances, could not causal contributions add so that two strong causal factors together determine the effect? And we might even ask why all causal factors should add similarly? Could there not be factors that work better with some than with others? That would hardly be a further deviation from the assumption that they are independent than is already contained in the dogma that they always add. If we believe in the three-way possibility there is an answer to these questions, but there seems to be little to say if we dismiss this possibility. The conclusion must be that Humphreys' model also in effect uses the result 7 – but that is a result that can be derived for probabilities, not necessarily for chances. And when derived for probabilities, an important assumption is that the effect *can* come about on account of one cause alone.

5.

The requirement of *sine qua non* figured extensively in the background to the argument above. This is a controversial requirement. But note that the argument does *not* build on the dogmatic assumption that *sine qua non* must hold, but on the effects of violating it in the case of probabilistic causality. Whenever there is a violation of the *sine qua non* condition for a probabilistic explanation, this is either an indication that there is more to be known, or that more than one ultimate probabilistic cause exists. In the first case, the explanation is clearly epistemically relativized and in the other, the argument above applies.

In fact it seems that a non-probabilistic causation can handle violations of the requirement of *sine qua non* better. If there were an instance where two deterministic mechanisms, separate from one another, are responsible for an event, we do not have the problem of explaining why their effectiveness is correlated: they are both always effective and hence they both explain the event. (Admittedly, this goes against many intuitions concerning deterministic causality, but those intuitions are not the issue here.)

It should also be noted that if the requirement of *sine qua non* in fact does hold even for probabilistic mechanisms, then the non-deductivist must admit that there exists, in principle, a method of deciding which one of two causal alternatives that was effective, since they cause slightly different effects. In that case, it would obviously be erroneous to cite both causal alternatives as the explanation; together they might cause one type of effect, and in isolation two other kinds. To go back to the example with operation Smoky, in this case it would be possible to study the kind of leukaemia in order to know if the witnessing of the detonation really was the cause. And unless we claim to have done that, we should not cite both alternative causes as part of the explanation, but merely as candidates. As long as a cause in a probabilistic law violates the requirement of *sine qua non* it cannot merit as anything more that a probable explanation of the effect.

References

Eells, E. (1991). *Probabilistic Causality*. Cambridge University Press, Cambridge.

Humphreys, P. (1989). *The Chances of Explanation*. Princeton University Press, New Jersey and Oxford.

Hållsten, H. (1999). Deductive chauvinism. *Synthese*, 120:49–59.

Hållsten, H. (2001). *Explanation and Deduction*. Almqvist & Wiksell, Stockholm.

Mellor, D. H. (1995). *The Facts of Causation*. Routledge, London.

Pitt, J., editor (1988). *The Theories of Explanation*, Oxford. Oxford University Press.

Rennie, J. (1996). Twelve major cancers. *Scientific American*, pages 92–98.

Råde, L. and Westergren, B. (1988). *Beta, Mathematics Handbook*. Studentlitteratur, Lund.

Salmon, W. (1980). Probabilistic causality. *Pacific Philosophical Quarterly*, 61:50–74.

Salmon, W. (1984). *Scientific Explanation and the Causal Structure of the World*. Princeton University Press, Princeton.

Salmon, W. (1988). Deductivism, visited and revisited. In Grünbaum, A. and Salmon, W., editors, *The Limitations of Deductivism*, pages 95–127, Berkeley/Los Angeles/London. University of California Press. (Reprinted in Salmon, 1998, pp. 142–165.).

Salmon, W. (1998). *Causality and Explanation*. Oxford University Press, New York and Oxford.

Skyrms, B. (1992). Probabilities, theories of. In Dancy, J. and Sosa, E., editors, *A Companion to Epistemology*, pages 374–378, Oxford. Basil Blackwell.

Suppes, P. (1970). *A Probabilistic Theory of Causation*. North-Holland Publishing, Amsterdam.

THE NATURE OF NATURAL LAWS

Lars-Göran Johansson

1. Introduction

Everyone who has done science in school has come across the notion of a natural law. Famous examples are the laws named after great scientists like Newton, Boyle and Coulomb. These are paradigmatic examples of scientific laws. The set of scientific laws also includes statements not usually referred to as laws, e.g. the *postulate* that the velocity of light is constant, the photo-electric *effect*, Pauli's exclusion *principle* and Schrödinger's *equation*.

However, we have no idea how to define the concept of scientific law. Many definitions have been discussed and rejected. It is astonishing that we can accept e.g. Schrödingers equation as a member of the family of scientific laws and still not agree on what constitutes membership.

A natural starting point in the discussion is to consider the logical form of a scientific law. Most philosophers agree that laws are true generalised conditionals, i.e. of the form 'for all x, if Ax, then Bx'. Some disagree, suggesting that scientific laws should be analysed as relations between universals, i.e., analysed in second order logic. I will discuss this idea in section 2, while for the time being assuming the stance of the majority.

It is immediately clear that being a universally generalised conditional (UGC) is not a sufficient criterion for being a law, because there are many true sentences which are UGC, which are not laws. Nelson Goodman's famous example in (Goodman, 1954) is 'All coins in my pocket are made of silver'. This sentence (or the proposition expressed) is not a law but an accidental generalisation. Assuming that a necessary criterion for being a law is that the logical form of the proposition is a generalised conditional, the problem of defining the concept of a scientific law is to find a criterion for distinguishing between laws and accidental generalisations.

At first sight this task does not seem too difficult. The accidental generalisations are most often of a restricted nature, concerning matters

J. Faye, P. Needham, U. Scheffler and M. Urchs, (eds.), Nature's Principles, 151–166.

of fact at a certain time or place. An accidental generalisation is not a true universality in the physical sense of the word, despite its logical form. However, Goodman (Goodman, 1946, Goodman, 1954) has shown how difficult it is to state a viable criterion using this intuition without ending in circularity. He also showed that an analysis in terms of contrary-to-fact conditionals provides no solution, because *their* truth conditions could only be stated in terms of whether they are supported by a scientific law or not. The circularity is obvious.

In a more recent work Bas van Fraassen uses the following example (van Fraassen, 1989), the type being discussed by several others, to show that the difference in generality is not the crucial point.

1 All spheres made of gold are less than 1 km in diameter.

2 All spheres made of ^{235}U are less than 1 km in diameter.

Let us for the sake of argument assume that both statements are true. Although both are universal (in the physical sense of the word) propositions we view 1 as an accidental generalisation and 2 as a law. The latter is a law because ^{235}U is fissile and the critical mass for a chain reaction is way below that of a 1 km diameter sphere. We have a strong feeling for the *physical necessity* of 2 and that is why it is not merely true but a law. In contrast, it appears as a purely contingent fact that gold does not exist in big lumps.

There is thus a strong connection between the concepts of natural law and physical necessity and it is tempting to say that laws are necessary truths of a certain type, viz. those expressing physical necessity. Then we face the task of explaining what we mean by physical necessity without using the concept of natural law. However, it is not easy; empiricists such as Hume, Quine and van Fraassen claim that the only acceptable notion of necessity is logical necessity and consequently physical necessity is an expression of bad metaphysics, not to say conceptual confusion. They have strong arguments on their side. My own view is that we could reverse the order of explanation: having first defined what we mean by a natural law we can say that a law is not merely true but expresses a physical necessity.

Connected to this stance is Hume's regularity theory of causation, according to which the necessity of cause-effect relations is just a feeling, viz. a matter of psychology. After having experienced a repeated sequence of the same type of events we are, just like Pavlov's dogs, conditioned to expect the second event when observing the first one. According to Hume this is all there is in the notion that the cause *necessitates* its effect.

Even if Hume is right in saying that we are conditioned to expect the effect to follow from its cause after a few observed sequences of the same sort, I still think that regularity theory is not sufficiently explanatory. I want to know *why* there are regularities in nature, not only to get a mere description of its psychological effects on me.

In this paper I will suggest a partly new analysis of the concept of scientific law. The core idea is that different types of laws require different analysis and that an important type, viz. those laws stating numerical relations between physical quantities, best should be seen as a kind of definitions. The idea has been expressed by Poincaré, but he corrupted it with a conventionalist view, which I do not find appealing. However, before developing my version of this idea I will give an overview of other ideas on the market using van Fraassen's penetrating discussion in his *Laws and Symmetries.*

2. Van Fraassen on scientific laws

Van Fraassen divides the current proposals about the concept of natural law into three main groups. The first one includes the theories of Pargetter, Vallentyne and McGall, all of which rely on a realistic view of possible worlds. The core idea is that a natural law is a proposition being true in all possible worlds. To avoid circularity it is then necessary to have a characterisation of possible worlds that does not depend on the notion of natural law. Pargetter, Vallentyne and McGall all take a strongly realistic view about possible worlds. But, the critics might reasonably ask: how do we identify these worlds? Obviously, we cannot use the usual way of characterising a possible world by saying that it is a world in which our scientific laws are valid. That would be circular. Another way of identifying these possible worlds is needed. However, we have not been presented with any and so we have not gained much.

The second view, first proposed by Mill, has recently been developed by David Lewis. The idea is that laws are the axioms in a future complete theory about the world. Not any axiomatisation will do of course; the resulting theory should be the simplest one entailing all observed empirical regularities. But there is a tension between simplicity and logical strength. Adding one more axiom to a certain theory T makes it stronger at the expense of being more complex. Thus we must somehow find a balance between simplicity and strength, and that is the problem, according to van Fraassen. In fact, we have three criteria for the choice: simplicity, strength and balance, and it is unclear how to select one particular axiomatisation out of a number of alternatives.

In passing it should be pointed out that Lewis does not use his well-known realism about possible worlds in his definition of the concept of natural law. Instead, he uses possible worlds to show that natural laws are necessarily true. Having first defined the concept of natural law, he defines those possible worlds which are accessible from our world as those in which our laws are valid, and then he defines necessary true as true in all accessible possible worlds.

The third view is proposed by among others Armstrong, Tooley and Dretske. They share the view that laws are relations between universals. This idea is in my view intuitively plausible; many laws are numerical relations between measurable quantities and these are naturally interpreted as a kind of universals. But what more precisely is the relation between universals, which would make it a fact that a particular object having a quantitative property logically implies that it has another property? Armstrong calls the relation 'necessitation': one (or several) universal necessitates another universal, symbolised in the form 'F-ness $\Rightarrow$ G-ness'. Armstrong then claims that this formula logically implies the usual law formulation $\forall x(Fx \rightarrow Gx)$, where '$\rightarrow$' signifies material implication. But how is that possible? How could there be a logical relation between these statements? The relation *necessitation* as obtaining between universals is a completely different relation than material implication, which is a relation between sentences. Armstrong sees the problem and claims that the solution is to identify necessitation among universals with material implication! This has not convinced the sceptics: As David Lewis ironically writes: "To call the relation 'necessitation' no more guarantees the inference than being called 'Armstrong' guarantees mighty biceps" (Quoted by van Fraassen, 1989, p. 98). In my view what is needed is a fleshed out theory which connects a second order logic of universals to first order predicate logic; one single axiom is not convincing.

Thus, even if we accept universals and relations among them in our ontology we have the inference problem: what kind of relation between universals guarantees that the facts instantiating these universals are related as antecedent and consequent in a material implication?

The prospects for all three ways of analysing laws are thus bleak according to van Fraassen and he concludes that there are no laws. This conclusion is however not unwelcome for him. It fits nicely into his constructive empiricism, according to which we can never know whether a theory is true or not, only know whether it is empirically adequate. A theory is empirically adequate if it has a model that is isomorphic with phenomena. It immediately follows that whether the theory contains any laws or not does not make any difference regarding empirical adequacy because a proposition p entails the same empirical consequences as the proposition 'it is a law that p'. Van Fraassen's conclusion is that we could just as well dismiss the concept of natural law as old-fashioned metaphysics.

However, his sharp distinction between the truth of a theory and its empirical adequacy cannot be consistently upheld, as I have argued elsewhere, (Johansson, 1996). It follows that laws might still have a role to play. It seems to me that any analysis of natural science that neglects, or explicitly rejects, the concept of scientific law is fundamentally wrong as a description of the conceptual structure of scientific thinking. People use the concept of natural law frequently when talking about science and it is not obvious that they misunderstand science when doing so. Well, van Fraassen might respond that scientists in fact do misunderstand themselves and it is the philosopher's task to help them to a better self-understanding. I disagree on the first point; scientists apply the concept of natural law, they have use for it and can use the concept in their discussions.

The discussion about laws is usually a dispute about ontology: Critics such as van Fraassen have claimed that there are no laws. I, on the other hand, are more interested in analysing the concept: what do scientists *mean* when they say e.g., that Maxwell's equations are the laws of electromagnetism? Although I'm very sceptical about the idea that meanings are a sort of abstract entities, I still think it profitable to consider how a certain expression is used in a set of contexts, which is a kind of semantic discussion. Thus, one can use pragmatic considerations in the analysis, because pragmatic components are involved in the meaning of concepts. This has not hitherto been done, or at least not to any considerable extent, in the discussion of laws. I will now propose some new ideas, but first give some preliminary distinctions.

3. Fundamental laws – derived laws

It is a common view among physicists and other natural scientists that some natural laws are more fundamental than others. For example, the principles of conservation of energy and the constancy of the velocity of light, Newton's laws and Maxwell's equations are all *treated* as fundamental, as the most basic principles in theoretical physics. They are not fundamental in the logical sense of being un-derivable; it is for example possible to derive Maxwell's equations from a set of other electromagnetic laws. Why then, are some treated as fundamental and not others? Perhaps the answer is that the scientist's sense of simplicity and elegance is the final reason; somehow scientists have agreed that one way of stating a theory is the most appealing one. Since my purpose is to take into account also pragmatic aspects of theory construction this feature should not be omitted.

New insights in a particular area of research often forces the scientists to restructure the theory in one way or another in order to accommodate the new findings. In so doing there is never only one option, but many. When deciding to restructure a theory we are always more reluctant to change certain things rather than others. An illuminating example is the fate of Newton's second law in the context of relativity theory. Instead of rejecting the law as inconsistent with relativity, it was upheld and the definition of mass was changed so that mass no longer is seen as a constant independent of frame of reference.

It is obvious that if some laws are treated as fundamental and some other as less fundamental, i.e., as derived, one uses the following inference principle:

P: if a statement has the logical form of a generalised conditional (or bi-conditional) and is derivable from a set of laws, it is itself a law.

Bas van Fraassen's example with spheres of ^{235}U illustrates the point. Statement 2 above is considered a law just because it can be derived from fundamental laws in nuclear physics and is a generalised conditional. Another example is the ideal gas law: it can be derived from the principle of energy conservation and some simplifying assumptions (which is the reason why it is about *ideal gases*).

If principle P is accepted we have reduced the problem of analysing the concept of natural law to the question what we mean by saying that these fundamental laws are natural laws.

4. Different types of fundamental laws

For my purpose it is suitable to divide the fundamental laws into three groups: i) relations between quantities; ii) conservation principles, iii) the rest.

Relations between quantities are the typical laws encountered in physics and chemistry textbooks. We have for example

- $F = ma$ — (Newton's second law)
- $\nabla E = \frac{\rho}{\varepsilon_0}$ — (Maxwell's first equation)
- $\frac{h^2}{2m}\nabla^2\psi - U\psi = -i\hbar\frac{\delta\psi}{\delta t}$ — (Schrödinger's equation)

The most well-known conservation principles concern the conservation of energy, momentum and angular momentum in any closed system.

Other, lesser well-known, are for example conservation of charge and baryon number.

The third category, the rest, is a heterogeneous ensemble containing for example the following principles:

- The constancy of the speed of light,
- The photo-electric effect,
- Pauli's exclusion principle,
- Quantisation of exchange of energy.

The reason for this division of the fundamental laws is as follows. The lawfullness of the laws in the first category I intend to explain as a kind of definitions. Those in the second category can be explained as a consequence of objectivity demands on the descriptions of nature. About those in the third category – the rest – I think it impossible to give any *general* explanation of their lawstatus; we have to treat them one by one. Perhaps some of the laws in this category will come out as consequences of other more fundamental principles when we understand more about nature. Thus my proposed analysis in this paper is at best partial and by no means complete.

5. Quantitative relations as definitions

Fundamental physics is a theoretical structure involving a lot of quantitative concepts such as mass, force, charge, electric field, magnetic field, energy, momentum, temperature etc. By agreement among scientists some are treated as fundamental. These agreements are codified in the SI-system of quantities and units. The international research community has agreed on seven fundamental quantities: length, mass, time, electric current, thermodynamic temperature, amount of substance and luminous intensity. All other physical quantities are introduced by implicit definitions in the form of quantitative relations between quantities often called laws. A simple example is Ohm's law, $U = R \cdot I$, which connects the potential difference U between two points in a circuit with the current I and the resistance R in the circuit. The electric current is a fundamental quantity and the potential difference is defined via other relations in terms of the work required to move a unit charge between the points. Work in turn is defined using force and force is defined by Newton's second law plus the law of gravitation (as we will see in a moment) and charge in terms of current and time. Then resistance, i. e., R, is defined via Ohm's law and several other laws as Mass $\cdot$ Length^2

· Time^{-3} · Current^{-2}. In other words, 1 ohm is 1 kg · m^{2} · second^{-3} · Ampere^{-2}.

It should be noted that it is possible to reverse the order of definitions, which in fact has been done some years ago in this particular case. After the discovery of the quantised Hall effect one has a very precise way of determining resistance and nowadays Ohm's law is used as a definition of potential difference instead of resistance. This possibility of reversal only underlines my point that Ohm's law is used as a definition. Improved possibilities of resistance measurements have not forced refinement or rejection of Ohm's law; it is still treated as an absolute truth.

One might object to the division of quantities into fundamental and derived as irrelevant to the analysis of laws, because being a collective decision it is subject to change. I accept the premise but not the conclusion; as already said, the present analysis is semantic, the purpose being to disclose what scientists mean when they talk about laws of nature. In this discussion the decisions made by scientists concerning quantities, units and measurements are relevant.

Ohm's law, however, is no fundamental law; it introduces a derived quantity. How about the possibility of viewing fundamental laws as implicit definitions of quantities? I will discuss classical mechanics and electromagnetism respectively as two important examples.

5.1 Classical mechanics

Classical mechanics divides into two parts, kinematics and dynamics. Kinematics concern only the positions and motions of physical objects, whereas dynamics account for the interaction between physical objects.

In order to describe positions and motions of physical objects we need basically only two scalar quantities, length (or distance) and time. Using these we can construct derivative quantities such as velocity, acceleration and position in space. Out of the seven fundamental quantities in the SI-system, kinematics thus needs only two, length and time. In the dynamical part of the theory, classically conceived, we have four fundamental laws, Newton's three laws and the law of gravitation. Two of these are fundamental laws relating quantities to each other, viz., Newton's second law and the law of gravitation. Newton's first law is a special case of the second one and can thus be omitted in this context. The third law states the reciprocity of interaction between two systems. So we have two fundamental laws left relating quantities to each other in classical mechanics:

- $F = ma$ Newton's second law

- $F = Gm_1m_2/r^2$ The law of gravitation

By restricting the application of these laws to systems where only gravitational forces are acting, the symbol F in the two equations refers to one and the same quantity. These laws contain four quantities: distance, acceleration, mass and force. Distance (or length) is a fundamental quantity whereas acceleration is a derived quantity in kinematics, defined by the formula

$$a = \frac{d^2s}{dt^2}$$

in which two fundamental quantities, distance and time, are presupposed. Finally, mass is a fundamental quantity, the mass of 1 kg defined as the mass of the international mass prototype. If we now view Newton's second law as the definition of force, assuming that we previously have defined mass and acceleration, the law of gravitation cannot be a definition but a strictly empirical law. But this seems arbitrary; we might just as well view the law of gravitation as a definition of force and the second law as empirical.

My view is that both these views are wrong, because they presuppose that the dynamic meaning of mass is previously known and well-defined. Mass is theoretical quantity and is introduced by these laws, just as force. One might think that the mass of an object is a simple observable and that the unit mass of 1 kg is given by ostentation; just point to the mass prototype in Paris. But, clearly, we cannot observe the mass of an object; just looking at the mass prototype does not tell us anything about its mass, not even that it has mass. Weight is another matter, but weight is not the same as mass; a piece of matter has weight because it has mass and is attracted by earth's gravitational attraction. Mass is a theoretical and dynamical property; to attribute mass to a physical object is to say that it has the *ability* to interact with other objects. My view is thus that mass and force both are *theoretical* quantities which simultaneously are introduced in mechanics in order to describe the interaction of bodies. The mass unit is given as the mass of the mass prototype, but the dynamical meaning of mass is given by the second law and the law of gravitation.

If we introduce two theoretical concepts we need two relations. Hence, Newton's second law and the law of gravitation simultaneously make up definitions of mass and force.

For a long time it was thought that the concept of mass occurring in the second law and in the law of gravitation respectively were different quantities, viz. the inertial mass and the gravitational mass, and for 200 years one was perplexed that they appeared to be equal. Nowadays we use insights from general relativity and understand that inertial mass and gravitational mass are in fact the same quantity. The argument is simple: by confining ourselves to a particular local system we cannot decide by any measurement whether the external forces on the system is due to gravitational attraction from another system or from its being accelerated. Hence the system's gravitational and inertial mass cannot be distinguished by empirical means and could just as well be considered as one and the same quantity. From these considerations one might conclude that a proper analysis of classical mechanics cannot be done in isolation, we need the background of the generalised relativity principle which says that all co-ordinate systems should be treated equal. It was stated in a restricted form (applied only to inertial systems with only mechanical forces) by Galileo and belongs in this restricted form to the back-ground for classical mechanics.

5.2 Electromagnetism

Electromagnetic theory cannot be constructed in isolation from mechanics. It presupposes the fundamental quantities time, length and mass and some of the derived quantities, most notably force and energy. Electromagnetism is the theory describing electromagnetic interaction between material bodies and in order to formulate the theory one needs to introduce three new quantities, viz. charge, electric field and magnetic field. (I neglect in this context the fact that the electric and magnetic fields can be defined in terms of the vector potential, which is the natural quantity to use in a relativistic description. My point here can be made without introducing the relativistic view-point, which therefore is neglected as unnecessarily complicating matters, although it simplifies the theory in a rather precise sense.)

The unit of electric charge is defined as the amount of charge passing a cross section of a circuit during 1 second when the current in the circuit is $1A$. It is thus defined in terms of current, which is one of the seven fundamental quantities in the SI-system. After having established the unit quantity of charge, one can easily define charge density. Then the vector quantities electric field ($\mathbf{E}$) and magnetic field ($\mathbf{B}$) are introduced by Maxwell's equations:

M1 $\nabla \cdot \mathbf{E} = \frac{\rho}{\varepsilon_0}$

M2 $\nabla \times \mathbf{E} + \frac{\partial B}{\partial t} = 0$

M3 $\nabla \cdot \mathbf{B} = 0$

M4 $c^2\nabla \times \mathbf{B} - \frac{\partial E}{\partial t} = -\frac{\mathbf{j}}{\varepsilon_0}$

These four differential equations together uniquely determine all the components of the two vector quantities electric field $\mathbf{E}$ and magnetic field $\mathbf{B}$, using charge density ρ and current $\mathbf{j}$ as independent quantities.

6. Definitions, yes, but what kind?

To say that a quantitative relation is a definition is not very precise, since the word 'definition' could mean many different things. To begin with, I want to make it clear that these definitions are not mere stipulations. For example, it is not just a matter of convention how we define matter and motion. Descartes defined the quantity of matter as extension, i.e., volume, and quantity of motion as volume times velocity. However, using these definitions he failed to construct a mechanical theory and the diagnosis of this failure is easy: his definitions were false.

My use of the word 'definition' in this context is more like Quine's according to which a definition is a part of theory which as a whole is true or false. I fully endorse Quine's stance that it is impossible to separate empirical and semantical aspects in the analysis of scientific theories.

The fundamental laws are definitions of theoretical predicates and together they make up a theoretical structure. It is related to empirical phenomena via the definitions of the fundamental quantities in the SI-system. If we are successful this theoretical structure 'fits' nature, i.e. the classification of phenomena, which is the content of using predicates, 'carves nature at its joints', to use a popular expression. A physical theory such as electromagnetism purports to be a true description of the world and the definitions of the central concepts are at best true or at least they have truth value. (If our theory is false they still have truth-value).

Saying that a theory purports to be true means at the very least that each sentence has truth-value. This stance, to endorse the principle of bivalence, is the minimal requirement for being a realist. For the core content of realism in a particular area of inquiry is the idea that what we are talking about exists independently of our thoughts and if we express our thoughts by making a statement, this statement is true or false depending on how the external world is. This is, by the way, precisely Dummett's view in his *Metaphysical Disputes over Realism* (Dummett, 1991).

Most realists want more: for example, those endorsing scientific realism claim that our best theories are in fact approximately true and the fundamental entities in those theories exist. Another version of realism is

Aristotelian realism about properties, which says that universals and in particular properties exist. In this case the Aristotelian would say that quantities such as mass, energy, and electric charge are universals and hence that those quantities exist as instantiated in the objects having those properties.

No doubt, people with realist inclinations, and that includes myself, regard these stronger forms of realism as better reflecting their deepest intuitions. However, both versions face problems which is not easily solved.

The debate about scientific realism has shown that scientific realists have problems of giving arguments for their stance which is not question-begging. Scientific realists justify their stance by using the mode of argument 'inference to the best explanation': the success of our scientific theories is best explained by the assumption that they are at least approximately true. Van Fraassen's reply (van Fraassen, 1980, pp. 19–23) is well-known: the best explanation of success is that those theories are empirically adequate, and that is possible even if they are false. Whether truth or empirical adequacy is the best argument for success depends obviously on what to count as the *best* explanation and that decision depends on whether you are a realist or not. As Dummett has pointed out, for example in his *What is a Theory of Meaning II?* (Dummett, 1993), a common neutral ground for the debate is needed. Dummett suggests that it is provided by turning to logic: do you or do you not accept the principle of bivalence?

Those who accept properties as constituents of the world face the problem of giving identity criteria for properties. As adherents to Aristotelian realism hold that two properties can be instantiated in exactly the same objects, it follows that properties cannot be identified with sets of objects. But what then is the criterion for identity; under what conditions are two properties A and B identical? No observation can provide a way of separating two coextensional properties and one might wonder whether we really need non-extensional properties.

It might be possible to give extensional identity criteria for quantities and if that is achieved I would be happy to accept quantities in my ontology. However, as long as that is not achieved I stick to a minimal form of realism, viz. to say no more than that theories are true or false independently of our states of mind. It is not necessary to include in the theory statements saying that one or several properties exist; to accept that e.g. Maxwell's equations are true does not force one to analyse them in second order predicate logic. They can be interpreted as shorthand expressions for statements about particular objects. For example, the first equation could be interpreted as "for every object it is the case that the divergence of the electric field emanating from it equals its charge."

Being reluctant to accept properties in my ontology, I prefer to say that the core content of laws, i.e., the equations, are relations between predicates, not relations between properties. Thus, the definitions of quantities should be seen as definition of predicates, of extensional linguistic entities with simple identity criteria: two predicates are identical if and only if exactly the same objects satisfy them.

In passing it could be mentioned that the expression 'physical quantity' is used in two different senses, as clearly indicated in Document IUPAP-25 (where the SI-system is codified): "One refers to the abstract metrological concept (e.g., length, mass temperature), the other to a specific example of that concept (an attribute of a specific object or system: diameter of steel cylinder, mass of the proton, critical temperature of water)." (I.U.P.A.P., 1987, p. 1) My use of the word 'quantity' in this context is that of the abstract metrological concept, i.e. in the sense of the type, not the token.

7. Completeness?

If it is accepted that there are a number of fundamental laws relating physical quantities to each other and if principle P is accepted so that general statements derivable from the fundamental laws also are laws, one might ask if this exhausts the category of quantitative relations? Let us suppose it doesn't; let us suppose that there is a purported law L in the form of a quantitative relation that cannot be derived from fundamental laws. This purported law L has gained good support by observations and no instance tells against it. Is this purported law really a law or merely an accidental generalisation which is such that as a matter of fact we have not hit upon counter instances so far?

The crucial thing is the character of the quantities involved. By assumption neither the new law candidate, nor its negation, can be derived from other laws. This could be so simply because the new law candidate contains a quantity which is not part of the so far accepted theory. There are now three possibilities. The first possibility is that the purported law is integrated into the rest of accepted science by redefinition or identification of quantities in the law with quantities being part of the accepted theory. There are some important cases of this in the history of science, the most important being the identification of temperature in a gas with mean kinetic energy. The second possibility is that sooner or later one will detect counter-instances, thus showing that the purported law was an accidental generalisation after all.

The third possibility is that the new law is accepted as a new fundamental one, an implicit theoretical definition of a new quantity.

The development of the ideal gas law is illustrative. The first step is Boyle's finding that the product of pressure and volume of a gas is a constant, when the temperature is constant. Next, Charles and Gay-Lussac found that under constant pressure the volume of a certain amount of gas is proportional to its temperature. (One might wonder how circular this statement is; the temperature was measured by a thermometer containing mercury, thus in effect measuring the increase of volume of mercury in the thermometer.) By joining these findings we get the ideal gas law, $pV = nRT$. (I neglect the introduction of the mole concept for the moment.) Much later it was proven that the ideal gas law follows from the energy principle, given a system of definitions, the identification of gases as collection of molecules and some simplifying assumptions.

How would we regard the ideal gas law if it proved impossible to integrate it into a theory starting from fundamental principles? There are reasons to think the researchers in the field would be reluctant to consider it as a law; it is not unreasonable to think that in such a case the observed regularity would be considered a local phenomenon, depending on unknown special circumstances.

Admittedly, I have not given a fully convincing argument for the claim that all laws in this category either are definitions of physical quantities or derivable from such definitions. A complete argument would require a systematic treatment of all laws in the entire domain of physics, a work far exceeding the space of a short paper. But I do think that the reflections given in this section provides argument strong enough to consider it worth while to embark on a more complete treatment.

8. Conservation principles

A category of laws of special interest are the conservation principles, because they are treated as fundamental by all scientists. Every conservation principle can be derived from a symmetry principle, according to Noether's theorem:

Noether's theorem: If a physical system is symmetric under a continuous parameter transformation, then there is a conserved quantity associated to that parameter.

Thus symmetry under time translations, space translations and space rotations result in conservation of energy, momentum and angular momentum respectively. There are a number of other more abstract parameter transformations under which we also require symmetry, such as phase transformation in electromagnetism which entails conservation of charge.

Why then should we require symmetry under a particular parameter transformation? Certainly, physicists do require that but why? Some would perhaps answer by referring to simplicity or beauty in the theory. However, there is, from an epistemological perspective, a better answer: symmetry requirements reflect a demand for objectivity in the description. Let us take symmetry under time translations as an example. This symmetry demand is the requirement that the form of our equations describing a particular system should be independent of any particular choice of zero point in time. In other words, we do not want the form of the description to be dependent upon the particular time when an individual experimenter think it suitable to start his clock. This requirement can be understood as a demand for maximum objectivity. This demand can be expressed as invariance under the transformation $t \rightarrow t + \Delta t$. Using the Lagrange formalism one can show that from this it follows that the conjugate quantity, the Hamiltonian, which expresses the energy of the system in question, is conserved. Noether's achievement was to give a proof that this can be generalised to all continuous parameter transformations.

My view is that the conservation principles belong to a second kind of natural laws. We regard them as fundamental laws because in the final analysis they are consequences of objectivity demands.

9. Summary

I have argued three things: i) that different types of laws should be given different analyses; ii) that laws expressing quantitative relations are all derivable from a set of fundamental laws which could be understood as implicit definitions of the involved predicates; iii) that conservation principles are in the final analysis consequences of objectivity demands on the descriptions of physical systems. It is obvious that there are a number of natural laws which neither can be seen as relations between physical quantities nor being conservation principles. About these laws I have presently nothing to say. However, I think I have covered a considerable proportion of the natural laws, at least a considerable proportion of known laws.

Natural laws are often said to be necessary truths. My analysis also provides a kind of explanation of this necessity. It seems reasonable to say that the necessity of quantitative relations can be understood as a kind of conceptual necessity. Why is e.g. Newton's second law a necessary truth? Well, it is part of the definition of the concepts mass and force; if we reject Newton's law we also change the meaning and extension of the predicates mass and force. Why is energy conservation a necessary truth; the answer is that it follows from an objectivity demand. As Quine (Quine, 1976, p. 69) observed, we often use the adjective 'necessary' when a consequence relation is shortened by omitting the antecedent. Thus we say 'it is necessary to pay your bills' which can be interpreted as 'if you want to avoid a law sue, pay your bills'. This explains the difference between a contingent accidental generalisation and a necessary law.

References

(1987). Symbols, units, nomenclature and fundamental constants in physics. *Physica* 146A, pp. 1-68. Document I.U.P.A.P.-25.

Dretske, F. (1977). Laws of nature. *Philosophy of Science*, 44:248–268.

Dummett, M. (1991). *The Logical Basis of Metaphysics*, chapter Metaphysical Disputes about Realism, pages 1–19. Harvard University Press, Cambridge, Mass.

Dummett, M. (1993). *The Seas of Language*, chapter What is a Theory of Meaning (II). Clarendon Press, Oxford.

Goodman, N. (1946). A query on confirmation. *Journal of Philosophy*, 44:383–385.

Goodman, N. (1954). *Fact, Fiction, and Forecast.* London.

Johansson, L. (1996). Van Fraassen's constructive empiricism – a critique. In Cohen, R., Hilpinen, R., and Renzong, editors, *Realism and Anti-Realism in the Philosophy of Science*, pages 339–341. Kluwer.

Lewis, D. (1973). *Counterfactuals.* Basil Blackwell, Oxford.

McCall, S. (1984). Counterfactuals based on real possible worlds. *Nŏus*, 18:463–477.

Pargetter, R. (1984). Laws and modal realism. *Philosophical Studies*, 46:335–347.

Quine, W.V.O. (1976). *The Ways of Paradox and other Essays*, chapter Necessary Truth, pages 68–76. Harvard University Press, Cambridge, Mass.

Tooley, M. (1997). The nature of law. *Canadian Journal of Philosophy*, 7:667–698.

Vallentyne, P. (1988). Explicating lawhood. *Philosophy of Science*, 55:598–613.

van Fraassen, B. C. (1980). *The Scientific Image.* Clarendon Press, Oxford.

van Fraassen, B. C. (1989). *Laws and Symmetry.* Clarendon Press, Oxford.

HOW THE *CETERIS PARIBUS* LAWS OF PHYSICS LIE

Geert Keil

After a brief survey of the literature on ceteris paribus clauses and ceteris paribus laws (1.), the problem of exceptions, which creates the need for cp laws, is discussed (2.). It emerges that the so-called skeptical view of laws of nature does not apply to laws of any kind whatever. Only *some* laws of physics are plagued with exceptions, not *the* laws (3.). Cp clauses promise a remedy, which has to be located among the further reactions to the skeptical view (4.). After inspecting various translations of the Latin term "ceteris paribus" (5.), the paper arrives at the conclusion that, on the most reasonable translation, there are no such things as cp laws, for reasons of logical form. Cp clauses have an indexical content, so that they need singular propositions as their habitat, not general ones. Cp clauses and the universal generalizations they are supposed to modify are not fit for each other (6.).

1. Cp Clauses and Cp Laws: A Survey

In the face of the "ragged character of the philosophical literature" on ceteris paribus clauses and ceteris paribus laws (Earman and Roberts, 1999, 461), I shall first give a brief survey of the field in order to background and locate my essential concerns. Let me distinguish five subtopics:

(i) Much of the recent literature on cp clauses and cp laws is motivated by a *skeptical attitude* regarding the laws of nature. Nancy Cartwright's influential view that "the laws of physics lie", in that they "do not tell what the objects in their domain do" (Cartwright, 1983, 55), has prompted a number of counter-reactions. One of these is the substitution of cp laws for strict laws, the idea being that universally quantified generalizations about empirical phenomena are plagued with exceptions, while cp clauses promise a remedy. Hedge a lawful statement with a cp clause, and it lies no longer. The unqualified law says something false, that is, it is subject to falsification by counterinstances. Then it gets

J. Faye, P. Needham, U. Scheffler and M. Urchs, (eds.), Nature's Principles, 167–200.

hedged by a cp clause, and as a cp law it says something true. Or so it seems.

(ii) It has often been observed that strict laws are hardly to be found in biology, sociology, or history. Or, for that matter, in economics: Long before the philosophy of science took an interest in the topic, cp clauses were widely used and discussed in economics (more widely used than discussed, actually).[1] It is true that "[m]uch work on the topic of provisos and *ceteris paribus* laws has been motivated by a concern to defend the special sciences" (Earman and Roberts, 1999, 472). Among those who feel the need to defend the scientific respectability of the special sciences against "physics chauvinism", many find comfort in the idea that physics might be no better off. They suspect that "science generally is riddled with ceteris paribus laws".[2] Some authors, including Cartwright, go so far as to hold that "ceteris paribus clauses are endemic even in our best physics" (Kincaid, 1996, 64; see also Morreau, 1999, 163). Let's call this view the *no-better-off* view. The no-better-off view need not come down to the skeptical perspective. Many advocates of the special sciences regard the cp-hood of the laws of physics as harmless, and merely press for equal treatment.[3] Others claim that this view is "based on a misguided egalitarianism about the sciences", while in fact it is "not '*ceteris paribus* all the way down' – *ceteris paribus* stops at the level of fundamental physics" (Earman and Roberts, 1999, 439 and 472). Hence, the debated issue is how widespread the anomalous phenomena are for which cp clauses promise to be a cure.

(iii) In particular, it is questioned whether the rough generalizations of *intentional psychology* are capable of scientific refinement. There has been a recent debate on cp laws in the philosophy of mind, stimulated by two seminal papers of Jerry Fodor's and Stephen Schiffer's,[4] and dealing with generalizations such as "If Tom wants a beer and believes there is one in the refrigerator, then he will go there to get it". The debate about the status of folk-psychological generalizations is older, of course. In Davidson's "anomalous monism", it's the "heteronomic" character

---

[1]For the history of the term "ceteris paribus" in economics, see Persky, 1990.

[2]Tye, 1992, 432. Likewise, Lakatos (Lakatos, 1978, 18) held that "*ceteris paribus* clauses are not exceptions, but the rule in science."

[3]"Nobody in his senses would hold that the laws of mechanics were invalidated if an experiment designed to illustrate them were interrupted by an earthquake. Yet [economics is continually critized] on grounds hardly less slender." (Robbins, 1984, 97–98) From this perspective, it seems more appropriate to say that the special sciences are *no worse off* than physics. Earman, Glymour and Mitchell (2002, 277) call the no worse off view "the CP defence of the scientific status of the special sciences".

[4]Cf. Fodor, 1991; Schiffer, 1991; Klee, 1992; Warfield, 1993; Horgan and Tienson, 1996, 115–141; Guarini, 2000.

of intentional generalizations that distinguishes intentional psychology from the rest of science. While "homonomic" laws are indefinitely refinable "by adding further provisos and conditions stated in the same general vocabulary as the original generalization", the heteronomic generalizations of intentional psychology can be improved "only by shifting to a different vocabulary" (Davidson, 1980a, 219).[5] In recent times, the debate has shifted from the more general question how widespread unstrict, cp generalizations are in science, to the more specific question whether the roles that cp qualifications play, and the ways of spelling them out, differ from psychology to physics. For example, it is held that cp clauses in intentional generalizations (e.g., "if there are no overriding desires") account for "same-level exceptions" only, while in physical laws, provisos are supposed to exclude all possible interferences.

(iv) In the philosophy of science, a major issue is whether cp laws can be saved from the charge of *vacuity*. Hedging a universally quantified conditional by a clause such as "unless something interferes" is often said to render the law-statement trivial. A cp law had better not be equivalent with the logical truth that all Fs are Gs unless they are not. One way of providing non-trivial truth conditions for cp laws is the "completer account": The cp clause indicates that some additional condition C exists that makes the antecedent of the law nomologically sufficient (Fodor, 1991, cf. Hausman, 1992, 133–139; Pietroski and Rey, 1995). If no such condition exists, the cp sentence in question is too vague so as to qualify as a cp *law*. Completer accounts go along with an interest in "distinguish[ing] legitimate from illegitimate uses of ineliminable *ceteris paribus* clauses" (Hausman, 1992, 133). It is often held that cp laws in the special sciences have completers in more basic sciences. Requiring, however, that some completer exist is one thing, while specifying it is quite another. As long as no independent specification of the completing clause is given, or a finite list of possible defeaters of the law, no real progress is made.

As the prospects for a breakthrough on this front are slight, some authors adopt a more radical strategy for saving cp laws from vacuity. Completer accounts, they say, leave the idea unchallenged that all genuine laws must be, or be amended until they are, universal generalizations. This idea is regarded as "fundamentally flawed" and "sharply at odds with standard scientific practice" (Woodward, 2000, 248 and 227). It is pointed out that "many a claim we believe to describe no reg-

[5]The conditions Davidson imposes on non-intentional laws are severe, and it has been objected that "if Davidson is correct, then there can be no purely physical laws either. ... Since there clearly are physical laws, Davidson cannot be correct" (Klee, 1992, 389).

ularity at all, nomological or accidental, we nevertheless accept as a law-statement" (Lange, 1993, 232). In a word, it seems that "we have many laws and few uniformities" (Cartwright, 1995b, 305). Consequently, these authors abandon the demand for exceptionlessness, and with that, regularity views of laws altogether. Alternative accounts include the views that laws of nature describe "Aristotelian natures", in the sense of "capacities" of natural substances (Cartwright, 1989 and Cartwright, 1992), physical forces (Lipton, 1999; Smith, 2002), or "causal powers" (Ellis, Chalmers), which are often regarded as dispositional properties of physical systems (Hüttemann, 1998), or as "defeasible causal tendencies" (Horgan and Tienson, 1996, Kincaid, 1996, 63–70).

(v) Still others dig deeper into the meaning of the term "ceteris paribus", trying to make quantificational sense of the clause, and trying to develop a semantics for cp claims.[6] Is the clause a quantifier?[7] If so, which variables does it bind? What is the logical form of cp laws? Or is the search for cp laws perhaps misguided, cp *sentences* not being lawlike? And how do we translate the Latin words "ceteris paribus" in the first place? Clearly, "other things being equal" is not the only option. In particular, the "ceteris absentibus"-reading introduced by Joseph deserves attention, according to which not the constancy, but the *absence* of other factors is required (Joseph, 1980). In the face of the different readings, it cannot be taken simply for granted that cp clauses are suited to modify lawful statements. Whether there are cp laws depends, among other things, on the logical form and the meaning of the cp clause.

2. The Need for Cp Laws: The Problem of Exceptions

The need for cp laws arises in view of the problem of exceptions, whereas the problem of exceptions arises only if laws are regularity claims. If laws, or lawful statements,[8] are conceived of as universal generalizations about empirical phenomena, they are false. Even our best candidates are plagued with counterinstances. The problem of exceptions was known and discussed long before Cartwright entered the fray. Mill was well

[6]Kurtzman, 1973; Joseph, 1980; Johansson, 1980; Silverberg, 1996; Morreau, 1999; Schurz, 2001a.

[7]"What is often apparently forgotten is that the phrase 'everything else' is analogous to a quantifier, which has the appearance of binding variables" (Kurtzman, 1973, 369).

[8]In the terminology I prefer, laws *are* statements, while the truth-makers for laws are empirical regularities or relations between universals, depending on the kind of law. Armstrong, Dretske, and Tooley use the term "laws" for the truth-makers, so that they need another word for the linguistic entities. "Law-statements" suggests itself. I stick to my nominalistic usage of "law", though, and call the worldly items "regularities".

aware of it, and so were Russell, Hempel, Lakatos and many others. Michael Scriven once opened a paper with the remark: "The most interesting fact about laws of nature is that they are virtually all known to be in error" (Scriven, 1961, 91).

Exceptions, or counterinstances, are cases in which the antecedent of the law is fulfilled while the phenomenon mentioned in the consequent fails to occur. In the case of laws of succession, for example, it can always happen that just at the moment the first event has taken place and the second event is due to occur, something interferes which prevents it. As Peter Geach puts it: "For every alleged uniformity is defeasible by something's interfering and preventing the effect, to assert the uniformity as a fact is to commit oneself to a rash judgment that such interference never has taken place and never will" (Geach, 1973, 102). According to the classical empiricist view of laws as describing exceptionless regularities, each case where the effect is prevented counts as a falsification of the law. "By 'laws' I mean descriptions of what regularly happens", says Cartwright (Cartwright, 1999, 4).[9] This view has much to recommend itself, for given that laws have the logical form of a universally quantified conditional, the characteristic of admitting no exceptions seems to be a built-in feature.[10] Still, it is not strictly analytic that universal generalizations describe exceptionless regularities, since some laws correlate items to which the notions of *regularity* and *exception* have no application, more of which below.

According to the skeptical view, no such universally quantified conditional about empirical phenomena is true. As Cartwright puts it, even the laws of physics "do not tell what the objects in their domain do. If we try to think of them in this way, they are simply false" (Cartwright, 1983, 55). "Indeed not only are there no exceptionless laws, but in fact our best candidates are known to fail" (46). As the quotations show, there are two different ways to express this result. Either you may say that the laws of physics lie, i.e. that they make false assertions, or you say that the laws in question *do not exist*. The difference is merely verbal, depending on your definition of "law". If a law is defined as a true lawful statement, speaking of false or lying laws is a *contradictio in adiecto*, just as speaking of true laws is pleonastic. Seen in this way, the wording that there are no exceptionless laws of physics is preferable,

---

[9]Lewis even *identifies* laws with regularities, claiming that "whatever else a law may be, it is at least an exceptionless regularity" (Lewis, 1986, 45). Substitute "describes" for "is", and you obtain the view under discussion.

[10]Armstrong introduced "oaken laws" which admit of exceptions, in addition to "iron laws" and "steel laws", which don't. According to Lewis, he would have better called the oaken laws "rubber laws".

the *no-laws* formulation being tantamount to the claim that there are no such things as false true statements.[11]

The core features of the classical empiricist view of laws are the following: Laws are true statements. Their logical form is that of a universally quantified conditional. They deal with empirical phenomena, not with the ideal objects of a model, nor with uninstantiated universals. They do not contain singular terms referring to particular objects, locations or times (the "Maxwell condition"). The conjunction of these conditions, most of which are indebted to logical empiricism, is sometimes called "the regularity account" of laws.

It has often been objected that all of these conditions can be met by accidental generalizations, as opposed to nomic ones. Some further condition seems to be needed, one that ensures the *modal force* of laws of nature, their *necessity*. The modal condition has been the source of much trouble. It has proven difficult to spell out exactly what the modal force consists of, and where it comes from. No analysis seemed reductive enough, for every proposal led into "the familiar unilluminating circle of analysis from a law's explanatory power, to its physical necessity, to its capacity for counterfactual support, to its lawlikeness, to its capacity to be inductively confirmed by its instances, to its explanatory power" (Lange, 1993, 243). For the present purpose, however, no modal reinforcement of empirical regularities is needed, since if Cartwright's version of the no-laws view is correct, there are no exceptionless regularities in the first place, regardless of their modal status. I mention this only because there is a second variety of skepticism about laws of nature, which is directed exclusively at their modal status. Bas van Fraassen "frankly advocate[s] the philosophical view that there are no laws of nature" (van Fraassen, 1989, 183), but his no-laws view, just like that of Ron Giere's, is in sharp contrast with Cartwright's no-*regularities* view, as his rhetorical question reveals: "If we say that the regularities are all there is, shall we be so badly off?" (ibid.). This kind of skepticism is nicely illustrated by van Fraassen's story of the omnipotent spaceship commander who travels through our galaxy and types the command "delete all laws" into his console, only to witness no changes whatsoever (cf. 1989, 90 f.). Van Fraassen envisages a world in which the deletion of all laws makes no difference to the course of events. What gets deleted are not the regularities, but only modal pseudo-facts about them.

[11]In order to avoid the harsh conclusion that there are no laws, Swartz distinguishes between "physical laws" and "scientific laws", and he makes it clear that "Scriven's point is about the pronouncements of science, about what scientists call 'laws'" (Swartz, 1985, 4).

In discussing the problem of exceptions, the modal status of the regularities can be neglected. So I shall stick to the regularity view, according to which any genuine law expresses "at least an absolutely unbroken regularity". This view leaves room for modal reinforcement, but is at odds with the view of laws as *relations between universals* and with the *dispositional* view. My reason for abiding with the regularity view is not, as some might suspect, that I like flogging a dead horse. The first reason is the intimate connection between laws and causality, while not just any law has a causal interpretation. The "principle of the nomological character of causality", as Davidson has called it, presupposes laws of a special kind. Secondly, and more important, not just any kind of laws calls for provisos. As mentioned above, the need for cp laws arises in the face of the problem of exceptions, and the problem of exceptions arises only if laws are claims of regularity. The dispositional account, by contrast, says that laws of nature describe dispositional properties of natural substances or of physical systems. According to this view, laws are "not about what things do but what it is in their nature to do" (Cartwright, 1992, 48), or "about the powers they possess" (Cartwright, 1983, 61). Laws do not state what *de facto* always happens, rather they say how a substance, given its nature, *would* behave under certain circumstances, e.g. in isolation. This conception of laws, however, is already a reaction to the intractability of the problem of exceptions. Statements about dispositions, powers, tendencies or natures, need no hedging, since such statements remain true even if the manifestation of the disposition is prevented by some external influence. This is why disposition talk is so useful: The possession of a dispositional property need not manifest itself in strict empirical regularities.

Nor are cp clauses needed if laws are relations between universals. Consider the much discussed case of interaction between gravitational force and electrical charge. "It is not true", says Cartwright, "that for *any* two bodies the force between them is given by the law of gravitation. Some bodies are charged bodies, and the force between them is not Gmm'/r^2" (Cartwright, 1983, 57). When forces interact, the law of gravitation does not give us the resulting force, and for that reason it does not tell the truth about how bodies in its domain always behave. But, no one should have expected this in the first place. Newton's law makes an assertion about the ratio between distance, masses and gravitational force, and this assertion is not falsified by charged bodies. The gravitational force between two charged bodies is still Gmm'/r^2, even if gravitation is not the only force present.

According to the concepion of laws as relations between universals, "laws eschew reference to the things that have length, charge, capacity [etc.] in order to talk about these quantities themselves and to describe *their* relationship to each other" (Dretske, 1977, 262). As long as these quantities are not *measured* quantities, but *properties*, i.e. universals, the relations between them are not affected by intervening factors or additional forces. Such laws simply don't make any claim about instantiations in our messy world. Assertions about relations between universals are not refuted by counterexamples to assertions about empirical regularities.[12] We need not enter, however, into a discussion regarding what the laws of physics essentially are. It will suffice, and seems more appropriate, to distinguish between different *kinds* of laws, as I shall do below. It is an exaggeration to maintain that *the* laws of physics lie. The claim under discussion is that *some* laws of physics are exception-ridden, and call for provisos.

3. How Some Laws of Physics Lie

Can the advocates of the regularity account provide any instances? Yes, some lawful statements about the physical world seem to describe what always de facto happens:

1 "Whenever a spark passes through a mixture of hydrogen and oxygen gas, the gases disappear and water is formed." (Nagel, 1961, 74)

2 "[S]tones thrown into water produce a series of expanding concentric ripples." (Nagel, 1961, 76)

3 "Whenever the temperature of a metal bar of length L_0 changes by ΔT, the bar's length changes by $\Delta L = k \times L_0 \times \Delta T$, where k is a constant characteristic of that metal." (Lange, 1993, 233)

These laws are cited by Nagel and Lange as examples of *causal* laws. Admittedly they are generalizations of the right kind, being *laws of succession*. They say something about which events always follow one

[12] Armstrong rejects uninstantiated universals, though. He claims that the relation of necessitation between F-ness and G-ness "*entails* the corresponding Humean or cosmic uniformity: $(x)(Fx \rightarrow Gx)$", so that "[a]ll genuine laws are instantiated laws" (Armstrong, 1983, 85 and 172). The entailment thesis is clearly wrong. In general, no relation between universals entails empirical facts about instances. Armstrong succumbs to what van Fraassen has described as "the inference problem", which can be expressed in the question: "What information does the statement that one property necessitates another give us about what happens and what things are like?" The answer is, from a logical point of view, "none whatsoever". There is an inferential gap. "Nothing less than a bare postulate will do, for there is no further connection between relations among universals and relations among their instances" (van Fraassen, 1989, 96 and 107).

another. Making assertions about the series of events, they have a direct causal interpretation, for causality is a relation between events. The view that only events can be causes has not gone unchallenged, but I shall assume it here without argument.[13]

The nomological view of causality says that every singular causal statement implies a strict law that covers the case. This view is held by Hume, Kant, Mill, Hempel, Popper, Stegmüller, Davidson and Lewis, among others. Davidson has called the view, somewhat clumsily, "the principle of the nomological character of causality" (Davidson, 1980a), or, more recently and less clumsily, the "cause-law thesis" (Davidson, 1995). In his version, the cause-law thesis says that if two events "*c* and *e* are related as cause and effect, there exist descriptions of *c* and *e* that instantiate a strict law" (Davidson, 1993, 312–313).[14] We need not know this strict law nor the descriptions, but they must perforce exist. According to the cause-law thesis, not just any connection between laws and causation will suffice. A causal law must exist that actually covers the case, causal laws being " laws that do subsume cause-effect pairs" (Cummins, 1983, 5).

The nature of causal laws depends on what kind of entities are admissible as causal relata. If it is taken for granted that only events can be causally related, then only laws about the succession of events can be causal laws, i.e. laws that say "Whenever an event of the type *c* occurs, it is followed by an event of the type *e*". Such laws make a claim about what happens. Not every physical law does so. For example, the law of the pendulum says something about the ratio of the length of a pendulum to its frequency of swinging. Boyle's Law says something about the ratio between the pressure, temperature and volume of an ideal gas. These laws are laws of *coexistence*, not laws of succession. They say something about simultaneous, or rather timeless, relations between properties, or between physical quantities.[15] The distinction between laws of succession and laws of coexistence goes back to Mill. "The phenomena of nature", Mill states, "exist in two distinct

---

[13]There are other candidates, for example Aristotelian substances (things and persons, in particular), facts, states, point events, dispositions, and tropes. All these entities have been promoted to causes and effects in various theories of causation. In my opinion, there are powerful arguments against all these candidates, but I cannot rehearse them here.

[14]Or, more precisely: "'*A* caused *B*' is true if and only if there are descriptions of *A* and *B* such that the sentence obtained by putting these descriptions for '*A*' and '*B*' in '*A* caused *B*' follows from a true causal law." (Davidson, 1980b, 16)

[15]More precisely, Boyle's law is "one of a large class of laws ... that are called 'constitutive equations' ..., which describe the behavior of specific materials". Such laws have been regarded as "*definitions* of materials" (Smith, 2002, 255). Cummins calls such laws "nomic attributions" (see below).

relations to one another; that of simultaneity, and that of succession", and accordingly they obey two kinds of laws (Mill, 1973, 323 [III, v, §1]). Thereupon, Mill makes a crucially important remark: "[U]nless there [are] laws of succession in our premises, there could be no truths of succession in our conclusions" (ibid., 325). Mill correctly states that no truth about what actually happens can be derived from truths about simultaneous relations alone.

This observation contradicts the standard account of the relationship between both kinds of truths. Laws of coexistence are typically expressed in the form of equations. Now it is often held that if such an equation contains a reference to time, it allows the *derivation* of an appropriate empirical truth, in the following way: "A differential equation containing time-derivatives, of whatever order, can in principle be integrated with respect to time, and will then tell us what later states of the system will regularly follow such-and-such earlier ones. [...] If functional laws could not thus be integrated with respect to time to yield actual *changes*, functional laws would be of little interest or use" (Mackie, 1980, 147 f.).[16]

This view, though mathematically compelling, is metaphysically mistaken. Integrating differential equations will never yield empirical truths about what *de facto* always happens. The only values that can be derived are those of *instantaneous* states of physical systems (sometimes called "point events"). Point events, however, must not be mistaken for empirical events of the kind mentioned in causal laws. The notion of an instantaneous state "is only a mathematical abstraction, which derives its entire meaning from the concept of a time interval" (Steiner, 1986, 251). Instantaneous states are inadmissible as causal relata, for causes and effects are *changes*, and every change takes time. Causation, on this picture, is a relation between *two changes*, not one between two temporally non-extended cross-sections of a physical process.[17]

The cause-law thesis assumes genuine laws of succession, i.e., laws that have cause-effect pairs as instances. Laws of coexistence, in con-

---

[16]Russell puts it like this: "[L]aws, as they occur in classical physics, are concerned with tendencies at an instant. What actually happens is to be inferred by taking the vector sum of all the tendencies at an instant, and then integrating to find out the result" (Russell, 1948, 316). Sheldon Smith has recently complained that many authors "fail to properly analyze the logic of the derivation of the differential equation". In particular, he emphasizes the difference between the solution of the differential equation, which describes the actual temporal behavior, and the law(s) used to *derive* the equation. Smith calls to mind the "Euler recipe" for deriving the equations of motion in point particle mechanics (Smith, 2002, 243 f.).

[17]The *two changes* view of the causal relation has been advocated by Ducasse (Ducasse, 1968). Incidentally, I use "event" and "change" interchangeably, just like Davidson does (see Davidson, 1995, 272). For a more detailed discussion of point events, differential equations and causal laws, see Keil, 2000, 249–260.

trast, are tailor-made to the view of laws as relations between universals. The no-laws view which *I* endorse is simply that no universally quantified conditional of the former kind is true. All laws of succession – asserting that whenever an event of the type c occurs, it is followed by an event of the type e – are falsified by counterexamples, or, if you like, by exceptions. In fact, the so-called "skepticism about laws" is an unfortunate turn of phrase. One does not require a distinctly skeptical attitude in order to notice that laws of succession which state empirical regularities are plagued with exceptions. It is a hardly disputable *empirical finding* that series of events do not exhibit exceptionless regularities. Cartwright's claim that the laws of physics lie is often discussed in too general terms, i.e. without paying attention to the kind of laws the skeptical view applies to. If properly delimited, the skeptical view should not be a contentious one at all. Cartwright draws an unclear distinction between "fundamental" laws and "phenomenological" laws. The all-important distinction, however, is of that between laws which state what actually happens and those which do not. In some places, Cartwright parallels both distinctions, and sometimes she says that "applying" the fundamental laws yields descriptions of the happenings: "If the fundamental laws are true, they should give a correct account of what happens when they are applied in specific circumstances. But they do not."(Cartwright, 1983, 13).

Consider Newton's law of gravitation. It does not say, nor does it imply, that whenever a body falls to the ground from a height of one meter, it will hit the ground at such-and-such a speed. Actually, most bodies will be slower, and some will never touch the ground, such as bread-crumbs which are caught by greedy seagulls. I have witnessed such cases. Did those seagulls falsify the law of gravitation? No. But the reason for their being innocuous is not that some kinds of disturbances could be neglected when testing succession laws. The reason is that Newton's law does nor purport to state which events always follow one another in the first place. It says something about how the gravitational force between two bodies depends on their masses and their distance. It does not say anything about the *total* force between actual bodies, nor does it mention other factors or disturbances. It asserts the existence of a gravitational force, and this assertion is not falsified by seagulls or other mischiefmakers.

The insight that many laws say nothing about the temporal behavior of physical systems has been used to counter Cartwright's bold claim that the laws of physics lie. The law of universal gravitation, so the counter-objection goes, "*cannot* misrepresent the motion of the body, because it says nothing specific about such temporal behavior"(Earman,

Roberts and Smith, 2002, 286; see also Smith, 2002, 245). Smith goes so far as to claim that "once one arrives at temporal claims within physics ..., one is generally trafficking in something other than a law" (Smith, 2002, 247).[18] But, recognising that some fundamental laws of physics say nothing about temporal successions is not easily distinguished from claiming that they do not describe the empirical facts, as Cartwright does. The only difference is that Earman, Roberts and Smith insist that fundamental laws do not even *purport* to describe empirical regularities. But this is more a matter of emphasis. The dispute seems to boil down to the question which expectations it is wise to entertain, and which amount of surprise is in order. Apart from that, I would maintain that Earman, Roberts and Smith keep to an unbalanced diet: The fact that force laws, such as the law of gravitation, do not purport to describe empirical regularities does not show that no laws whatsoever do.

It has been noticed before that "[s]cientists do not try to describe natural events in terms of what always happens" (Geach, 1973, 102). This observation has prompted different reactions, though. Geach draws an Aristotelian moral from the failure of physics to provide strict laws of succession. Nature does harbour invariances, yet her stable and general traits are not to be found in strict empirical regularities but in the essential properties of natural substances. Sometimes these properties do manifest themselves in empirical regularities (especially in the laboratory), but sometimes not, due to some interference or to abnormal conditions. *Hydro-chloric acid dissolves zinc* – this is a general truth about the natures of hydro-chloric acid and of zinc. Using Cummins' terminology, we may call these general truths *nomic attributions*, i.e. "lawlike statements to the effect that all X's have a certain property P" (Cummins, 1983, 7). We may further say, as Horgan and Tienson do, that statements like "Hydro-chloric acid dissolves zinc" report "defeasible causal tendencies", and as such, are undoubtedly general truths with empirical content. "Since there are systematic patterns of defeasible causal tendencies," Horgan and Tienson maintain, "there should be soft laws reporting such tendencies" (Horgan and Tienson, 1996, 121).[19]

[18] See also Earman, Roberts and Smith, who hold that "differential equations with time as the independent variable describing the evolution in the physical magnitudes of a given system" are not laws at all, but applications of a theory to a specific case. "Differential equations of evolution type" cannot count as laws because they are derived using non-nomic boundary conditions of specific cases (Earman, Roberts and Smith, 2002, 298 and 286).

[19] Lipton even submits that describing dispositions is precisely the job of *ceteris paribus* laws. In his view, cp laws "describe dispositions or forces that are stably present whether or not all things are equal" (Lipton, 1999, 155). The moral he draws from his dispositional account is that cp laws "point to the simpler reality that sometimes underlies the complexity of the phenomena" (ibid., 163).

The fact that these systematic patterns fall short of strict regularities does not make them any less empirical. I take it for granted that Horgan and Tienson have *occurrent* patterns in mind. Otherwise, it would be hard to see what exactly "patterns of tendencies" should be, ontologically speaking. Tendencies qua dispositional properties can hardly be said to build patterns unless they get instantiated. But as to the "soft laws" reporting the patterns of manifestations, disturbances stemming from other tendencies do not need to be ruled out.[20]

Working scientists are much more Aristotelian in spirit than empiricist philosophers of science often tried to make us believe. The return of Aristotelian natures, essential properties and dispositions is by no means a relapse into prescientific scholasticism, for "many of the most pressing and puzzling scientific questions are questions about properties, not about changes" (Cummins, 1983, 15). Why did the litmus tincture turn red? If the triggering cause of the event is sought for, the answer would be: because acid was poured into the test-tube. Citing the causing event, however, will fail to enlighten anyone who has witnessed the experiment. In science, we seldom ask for the triggering cause. At least in observable cases, the cause is obvious enough. More often we want to know something about the dispositional properties of the substances involved, trying to find out by what exactly the effect produced was F, rather than G.

Physical science is doing quite well without strict laws of succession. The fundamental laws that physicists are so proud of are different in character. The only ones who are left empty-handed are the philosophical champions of the cause-law thesis. The good news is that only some laws of physics lie; the bad news is that these are exactly those laws which Kant, Mill, Hempel, Stegmüller, Davidson and Lewis invoke, and which they need, claiming that whenever two events are related as cause and effect, there exists a strict empirical law that covers the case. Laws about simultaneous, or rather timeless, relations between universals, may be strictly true, but have no causal interpretation. With generalizations about tendencies or dispositions, it's more complicated, but it should

[20] As Horgan and Tienson say, "it is no part of the law's role to delimit all ways that a causal tendency belonging to this pattern might get defeated when interacting with other defeasible causal tendencies" (ibid., 122). Hempel makes a similar point: "A scientific theory propounds an account of certain kinds of empirical phenomena, but it does not pronounce on what other kinds there are. The theory of gravitation neither asserts nor denies the existence of nongravitational forces, and it offers no means of characterizing or distinguishing them." (Hempel, 1988, 30) Earman and Roberts suggest that the provisos Hempel speaks of "are not provisos proper but are simply conditions of application of a theory" (Earman and Roberts, 1999, 444).

have become clear that they do not support the Davidsonian cause-law thesis either.[21]

The scientific significance of strict laws of succession has always been overestimated by D-N theorists. And this is why the achievements of physics are not belittled by the verdict that some laws of physics lie. This is also why I am not intimidated by the warning that "if we demand that all genuine laws must be exceptionless, it follows that we know very few laws" (Woodward, 2000, 228). I do not demand that all laws must be exceptionless, only that empirical laws of succession be. The notion of exception has no direct application to laws about properties, exceptions being counter*instances*. Remember that assertions about relations between universals cannot be refuted by cases which falsify succession laws.

I distinguished succession laws from coexistence laws, following Mill. I am well aware that the latter category is a very mixed bag, including force laws, conservation principles, functional laws, nomic attributions, composition laws and some others. A comprehensive classification of physical laws is still to come, but for the present purpose, my rough division will do.

4. *Yes, But* Reactions to the Skeptical View

I said above that the no-laws view, properly delimited, should go undisputed, and that it is an empirical finding rather than a philosophical claim. This was of course an exaggeration. The view in question is not undisputed. The philosophy of science is bristling with counter-reactions, with "Yes, but" replies. The skeptical view is easily put forward, and often enough the friends of the laws even buy into it, calling it a superficial insight, or one that rattles an unlocked door. The exciting job is to parry all the *yes, but* reactions that will inevitably follow. I have tried to do this in some detail elsewhere.[22] Let me just list the

[21]Some philosophers claim that these *are* causal laws, in that "they cite real causes at work", since tendencies "are partial causes" (Kincaid, 1996, 65). Obviously opinions are divided concerning what causal laws are about. I reserve the term "cause-law thesis" for the kind of laws described above: universally quantified conditionals which subsume cause-effect pairs, in the straightforward way. In this view, a cause is a causing event. General traits, factors, forces, or tendencies may be referred to in causal explanations, but they cannot cause anything. There are no such things as "general causes". There are general causal *sentences*, like "Smoking causes cancer", but these are deceptive. This alleged generalization describes a disposition, or a tendency, or perhaps a statistical fact, but it is not a universal truth concerning events. It does not allow deduction of singular causal truths. It is not the kind of law that the Davidsonian cause-law requires, and which can be used in D-N explanations of the occurrence of events.

[22]Cf. Keil, 2000, 182-260.

most common replies the friends of the laws have in their quiver, before discussing the appeal to cp clauses.

Some say that empirical laws are at least *approximately* true. Others say that laws are *idealizations*: they don't deal with empirical objects, but with the ideal objects in a model, and they are entirely true of those objects. This view is reminiscent of the Platonist-Galilean view that the Book of Nature is written in mathematicalese. Still others say that the empirical laws which are true to the facts are *statistical* laws which cannot be falsified, strictly speaking, by single instances. Some philosophers of science promote *instrumentalism*. Laws, they say, are useful instruments in the scientific enterprise, but they need not be true to serve that purpose. Cartwright sometimes flirts with this line of thought. "We are lucky that we can organize phenomena at all", she says. "There is no reason to think that the principles that best organize will be true, nor that the principles that are true will organize much" (Cartwright, 1983, 53). Still another remedy is the *dispositional* view of laws mentioned above. And last but not least, it is said that empirical laws are false if taken at face value without qualification, but true if qualified with a cp clause.

We may try to systematize these reactions to the skeptical view by relating them to the various features of lawfulness which the philosophy of science has come up with. Taking for granted the classical empiricist view, and focusing on the three conditions (i) that laws have to be true, (ii) that they deal with empirical phenomena, and (iii) that they are strict – i.e. admit of no exceptions –, we can say that most of the proposals just enumerated amount to dropping one of these conditions in favour of the others. For example, the instrumentalist says: Laws are strict and deal with empirical phenomena, but they need not be true. The Platonic idealist says: Laws are true and strict, but not true of the empirical world. The champion of cp laws says: Laws are true and about the empirical world, but they are not strict.

Calling counterinstances to empirical laws "exceptions" is tendentious. It's a kind of euphemism. One could as well speak of falsifying instances and leave it at that. While no special skeptical attitude is needed to notice the problem of exceptions, a widespread *charitable* attitude explains why we do not go around decrying all laws as false, even in the face of counterinstances.[23] Some counterinstances, the friends of the laws hope, can be explained, or explained away. Such cases are

[23] "We do not go around decrying all physical laws as false (although they are not exactly true) because they serve the crucial explanatory role of singling out a preferred value from which, it is alleged, all deviations can be explained." (Scriven, 1959, 467)

reckoned as *pseudo*-falsifications.[24] Prima facie, they falsify the law in question, but for some reason or other we think that we can eventually cope with them. If we expect to cope with the counterinstances in a certain way, we call them "exceptions", which are due to "disturbing factors". If we expect to cope with them in a different way, we call them "inaccuracies". Pietroski and Rey have introduced a very useful distinction between catastrophic and noncatastrophic interferences (Pietroski and Rey, 1995, 94–97). The latter are distinguished by the fact that the law interfered with yields at least approximately the correct result. The inaccuracy of the prediction is due to an imperfect realization of the conditions stated in or presupposed by the law. The mercury in the thermometer is impure, the vacuum is incomplete, friction occurs etc. But a vacuum can be increased gradually, and if the measured result correspondingly approaches the predicted result, the experimenter feels entitled to close the remaining gap by Galilean idealization. Idealizations, it is said, "involve exaggerating some actual property toward some limit" (Hausman, 1992, 131).

With the behaviour of greedy seagulls, it's different. Such catastrophic interferences are not amenable to systematic consideration, and they can prevent that something even remotely similar to the predicted effect occurs. I wish to suggest that it are just these catastrophic interferences which create the need for cp clauses, while inaccuracies and impurities call for idealizations and approximations. If ideal conditions are insufficiently realized, "the problem is not in saying precisely what is involved in the idealization but in relating it to the real world which is not ideal" (Earman and Roberts, 1999, 457). With catastrophic interferences, it's different. They call for an *unspecified* proviso, for there is no reason to suppose that a complete list of such incidents is available for any empirical law. Cp clauses proper are needed when the *cetera* are not known.

Since real world situations are susceptible to both catastrophic and noncatastrophic interferences, both must be made provision for. Even if, in a given law, all the impurities in the explicitly stated conditions are accommodated by appropriate idealizations, there is still a proviso needed for catastrophic interferences. Some provisos are even dual-purpose tools. Unspecified *ceteris absentibus* clauses (of which more

[24] "[S]cience needs a concept of pseudo-falsification because a countervailing cause or interfering agent may be at work generating the 'counterinstance'. It is only under closed conditions ... that a theory can be given a fair test or that a crucial experiment ... becomes possible" (Bhaskar, 1978, 161). "We call the laws by the honorific term 'idealization' rather than by the pejorative term 'falsehood' because ... we can easily perform thought-experiments in which the disturbing factors ... are removed." (Joseph, 1980, 774)

below) are strong enough to rule out both kinds of interference with a single blow.

Let us now look into the question what exactly cp clauses do with the statements they qualify. Which abuses they are supposed to redress is clear enough. We are talking about provisos "without which a putative law would not be a law ... for the fundamental reason that it would be false unless qualified" (Earman and Roberts, 1999, 444). The suggestion is that cp clauses are capable of turning false generalizations into true ones.

Cartwright sums up her early views as follows: "[T]he fundamental laws of physics do not describe true facts about reality. Rendered as descriptions of facts, they are false; amended to be true, they lose their fundamental, explanatory force" (Cartwright, 1983, 54). Leaving aside the issue of explanatory force for the moment, the question is how a false generalization can be "amended to be true". It could be done by modifying the law so that it no longer applies to the irregular case. In the face of a counterinstance, we can identify the interfering factor and use this information to improve the law. We simply complete the antecedent with a clause that excludes that very factor; for example, "no friction occurs", or "no electromagnetic forces are present".[25] Of course, such an ad hoc amendment will restrict the range of application of the law to cases where the factor in question really *is* absent. In some cases, we can do even better. Friction, for example, is sufficiently understood as to admit of systematic consideration in a super law, that is, in a combination of two laws by means of vector addition. Physics textbooks record the law of falling bodies with friction in a medium as a familiar example.

Amending a law to be true in either of the two ways has a price, though. As Cartwright observes, "a law that holds only in restricted circumstances can explain only in those circumstances" (Cartwright, 1983, 155). It goes without saying that disturbing factors cannot be removed by simply postulating that they are absent. Disturbed cases may still occur that will remain unexplained by the specified law. Not only is there no reason to suppose that a complete list of possible catastrophic interferences will be available, it is not even advisable to make provisions for every conceivable incident. If we keep making ad hoc amendments to rule out various kinds of disturbances, we keep narrowing down the

[25]Often, such completions are framed as "boundary conditions", in order not to burden the law itself, or the scientific theory it belongs to, with too many provisos. Boundary conditions are mostly conceived as restricting the *range of application* of the law, which has to be done before its truth-value can be judged.

range of application of the law. Carried to extremes, we will end up with a true conditional with a hypercomplex antecedent that is fulfilled only once in the lifetime of the universe. Cartwright draws just this moral: "If I am right, a law that actually covered any specific case, without much change or correction, would be so specific that it would not be likely to work elsewhere" (Cartwright, 1983, 112).[26] There is, in other words, "a trade-off between truth and explanatory power" (ibid., 59). In the world we inhabit, you cannot have both at once – statements which have a wide range of application *and* which tell the truth about what happens in every specific case. "The truth doesn't explain much", Cartwright summarizes. That this tension exists is no a priori insight. If the world were not such a messy place, both ends could be achieved with statements of the same kind. It is obvious that the availability of laws which are both true and explanatory was an assumption implicit in the D-N-model of scientific explanation. This assumption was mistaken, and this fact contributed to the decline of the D-N model.

Catastrophic interferences can destroy laboratories, but they do not worry theoretical scientists excessively. Shielding an experimental set-up from catastrophic disturbances is an engineering task rather than a matter of hedging the relevant law(s) with countless ad hoc clauses. In fact, "physicists do not add the required clauses; they leave the expression of the law as it is" (Smith, 2002, 240).

And worse yet, even if a complete list of possible catastrophic interferences were available, this list could arguably not be used to turn a false nomic generalization into a true one. For if we had such a list, and used it to complete the antecedent of the law, we would still "not have a strict law, because the completed antecedent would yield a universal conditional that is true but not a law, only an *accidental generalization*" (Lipton, 1999, 160) The completed antecedent would specify the cases where the interfering actors are absent, but since it is generally a contingent matter when and where interference occurs, such "a strict antecedent purely in occurrent terms would fail to be lawlike", according to Lipton (ibid., 165).[27] I'm not quite sure whether this claim of

[26]This consequence had been discussed some decades before in the controversy between Hempel and Dray about laws in history. Scriven had pointed out then that a certain historical 'law' with a couple of *ad hoc* amendments "has become more trivial, i.e., less general, in the course of becoming more nearly correct and now appears quite possibly to have only one instance" (Scriven, 1959, 455).

[27]Similarly, Earman, Roberts and Smith observe that in the case of interaction between gravitation and electrical charge, hedging the law of gravitation with the clause "no other forces present" would deprive the resulting statement of its lawful character, since the proviso would bring in a "non-nomic assumption" (Earman, Roberts and Smith, 2002, 287; cf. Smith, 2002, 247 f.).

Lipton's is correct, because the demand that the law be nomic does not apply to its antecedent clause, strictly speaking. Given that laws are conditionals, it seems that the antecedent does not *maintain* that none of the interferences is in play, contrary to what Lipton suggests (ibid., 160). Rather the completed law says that *if* all these specified factors are absent, such-and-such will happen. As far as I can see, such a conditional could still be lawlike, provided it does not contain singular terms referring to particular objects, locations or times. So I think that the better way to frame this kind of objection is the one taken above: We would end up with a law with a hypercomplex antecedent which applies only once, viz. to the particular situation for which it was tailored. Formally, the sentence could be lawlike, but it would fail to be a *covering* law. But, all this holds only if we had a *complete* list of what could possibly disturb a real world situation, and there is no reason to suppose that knowledge of such a list is within our cognitive powers.

5. How to Translate "Ceteris Paribus"

Not just any qualification which aims to restore the truth of a law qualifies as a cp clause. It is high time to clarify the Latin phrase "ceteris paribus", before we can hope to determine which role cp clauses can play in laws. There is little reason to switch, midst of the formulation of a law, to Latin and return to English after two words. So, how to translate "ceteris paribus"? As far as I can see, there are at least six standard readings, only some of which can be judged translations:

1 other things being equal

2 all other things being equal

3 under normal conditions

4 under ideal, or optimal, conditions

5 provided nothing interferes

6 all other forces being absent (*ceteris absentibus*)

The difference between clause (1) and clause (2) is that the first leaves it open how many of the other things are supposed to be equal. Clause (2) invites the comment that *all* other things are never equal. It is very unlikely that two events ever occur under exactly the same circumstances – that is, under circumstances of exactly the same kind, apart from spatiotemporal position, of course. As far as we know, history does not repeat itself. (The issue of eternal recurrence I set aside, for in this case, the very notions of repetition, and of numerical identity, become doubtful.) It was one of Russell's arguments against causal laws that the more precisely the antecedent of a causal law becomes specified, the less likely it is that such a case will ever recur.[28] All circumstances never being equal, hedging a law by clause (2) is cognate with the idealization strategy. Such a "law" had better be expressed in the subjunctive mood: Such-and-such *would* happen if, *per impossibile*, all other things *were* equal. Lipton has dubbed this problem "the *problem of instantiation.* Many cp laws appear to have no instances at all, because things are never 'equal' in the requisite respect" (Lipton, 1999, 157).

If, on the other hand, the proviso fixes less than all other things, as in clause (1), the natural question is: which ones? The reply is that we do not know. If we did know, we could use this knowledge to specify the antecedent of the law, as described above. We would simply rule out the circumstances in which the effect does not materialize, or we would combine all known factors or forces to formulate a super law. Such a specification, however, should not count as a cp qualification. We should reserve the term "cp clause" for unspecified provisos, in order to mark the difference between both manoeuvers. It has often been observed that cp clauses "are needed in science precisely when it is *not* clear what the 'other things' are" (Pietroski and Rey, 1995, 87), that is, when "no definite claim is in the offing" (Earman and Roberts, 1999, 452).

Cartwright calls a law that holds only in specified circumstances a cp law as well (Cartwright, 1983, 47). Doing this, she confounds both devices, and covers up the reasons why an unspecified proviso is still needed after we have done our best to improve the law by specifying the antecedent. The first reason is that the need to make an indefinite number of specifications makes it impossible to spell out the antecedent, so that the law would be "incapable of being written down explicitly" (Giere, 1988, 40).[29] A second reason is that making provisions for a certain disturbing factor will not even do for that very factor. Consider again the law of thermal expansion as applied to metal bars: "Whenever the temperature of a metal bar changes by ΔT, the bar's length changes by ΔL". This unqualified generalization does not seem to be true. We

[28] "As soon as the antecedents have been given sufficiently fully to enable the consequent to be calculated with some exactitude, the antecedents have become so complicated that it is very unlikely they will ever recur" (Russell, 1986, 188).

[29] See also (Lange, 1993, 240). Armstrong seems prepared to bite this bullet, holding that it "could even be that the statement of, say, Newton's first law as an iron law, would have to be of infinite length" (Armstrong, 1983, 149). "Iron laws" is his term for exceptionless laws, while he calls cp laws "oaken": "Unlike iron laws, oaken laws do not hold *no matter* what. They hold only in the absence of interfering factors" (ibid., 106).

may consider making the amendment "unless the bar is hammered inward at one end" (Lange, 1993, 233). But, this amendment would be of dubious value, because it is both too weak and too strong to restore the truth of the law. Not only does it fail to exclude other defeaters, it also wrongly excludes ineffective cases of hammering. After all, "the bar may be hammered upon so softly and be on such a frictionless surface that the hammering produces translation rather than compression of the bar" (ibid., 235). Hence, simply forbidding hammering will not do. Both considerations indicate that specified clauses do not capture the intuitive sense of "ceteris paribus", and both corroborate the insight gained in the precedeing section: As a general remedy for the problem of exceptions, the strategy of specifying antecedents is hopeless. And in fact, serious science does not employ it.

If all other things are never equal, "strictly speaking no *ceteris paribus* law literally applies" (Kincaid, 1996, 64). Kincaid concludes that cp laws "are apparently false when other things are not equal" (ibid. 67). But this conclusion comes too quick, for reasoning from a law's inapplicability to its falsity is invalid. If we think of the clause as a part of the antecedent, a cp law is not ipso facto false should the cp clause not be fulfilled. Just like Lipton, Kincaid seems to assume that a cp law *asserts* that all things are equal. But, under the assumption that a cp clause is an additional if-clause, laws qualified with such a clause assert what always happens *if* everything else is equal. If the proviso is not fulfilled, the conditional is not false, but vacuously true. – But all of this holds only if we choose clause (2) instead of (1), and this I would not advise.

The remaining clauses can hardly be reckoned literal translations of the Latin phrase "ceteris paribus". Yet it has become customary to discuss them under that heading, since there is some functional equivalence to the "original" cp clause. Clause (3), "under normal conditions", prompts the question which conditions count as normal. There are two options here, a statistical notion of normality and a teleological one. The statistical reading amounts to treating the normality clause as a quantifier. In this case, a cp law would be "a crude statistical law: *for the most part ...*" (Cartwright, 1983, 47). But most authors agree that the notion of normality invoked in clause (3) is not statistical.[30] The reason is obvious: The circumstances under which the consequent of an empirical law becomes exactly true may be rare. As for the counterin-

[30]Cf. Cartwright, 1983, 47; Pietroski and Rey, 1995, 84 f.; Silverberg, 1996, 216 f.; with the exception of Schurz, 2001a, who bases his statistical notion of "normic laws" on a generalized theory of evolution. Schurz confines his account, however, to the phenomenological "system laws" of self-regulatory systems, in contrast to the "laws of nature", as he calls the fundamental laws of physics which are not restricted to special entities.

stances due to the imperfect realization of ideal conditions, the rareness of positive instances of the law is only natural, for the probability of *ideal* conditions (no friction, a perfect vacuum etc.) is zero. As Lipton puts it: "Many cp laws have no instances, and it cannot be the case that most Fs are G if none are" (Lipton, 1999, 159).

Hempel has made it clear why probabilistic construals and cp laws are tailored to cope with different kinds of counterinstances. It is tempting to raise the empirical adequacy of, say, the theory of magnetism by a probabilistic construal, i.e., by resorting to laws such as "Given that a metal bar is magnetic, the probability that iron filings will cling to it is p_2". Hempel dismisses this idea, vaguely suggested by Carnap, on the grounds that "surely, the theory of magnetism contains no sentences of this kind; it is a matter quite beyond the theory's scope to state how frequently air currents, further magnetic fields, or other factors will interfere with the effect in question" (Hempel, 1988, 25). Thinking otherwise is asking too much of scientific theories. Nobody knows which numerical value of probability to assign to catastrophic interferences, i.e. to the chance that one out of an indefinite number of potential interferences will occur. Hempel concludes that "a probabilistic construal cannot avoid the need for provisos" (ibid.).

If the notion of normality invoked is not statistical but teleological, we arrive at clause (4), "under ideal conditions", or "under optimal conditions". Variants of the teleological clause include "other things being *right*" (Cartwright, 1983, 45), "in favourable circumstances", and "if the relevant circumstances do not change". Such clauses raise the question: "Relevant to, or optimal for what?" One obvious answer is, "for the effect to occur". This seems to be the only answer which makes sense, but it incurs the charges of circularity and vacuity. The teleological reading is responsible for the bad reputation of cp clauses in the empiricist camp.[31] In effect, teleological normality clauses are equivalent with clause (5), "provided nothing interferes". As long as no independent characterization of the interfering factor can be given, advocates of clause (5) will have to explain what distinguishes the generalization "*B*s follow *A*s provided nothing interferes" from the statement "*B*s follow *A*s unless they don't".

Since cp laws had better not be equivalent with logical truths, clauses (4) and (5) are often combined with completer accounts of cp laws.

[31] According to Popper, clauses that fix circumstances "will be the more interesting the more specific they are, and the more testable they render the original theory. I therefore suggest that *ceteris paribus* clauses should be avoided and, more especially, that they should not be imported into the discussion of the methodology of the natural sciences" (Popper, 1974, 1187).

The proviso "unless something interferes" is regarded as an invitation to determine the disturbing factor, whereas, as Geach puts it, "a vacuous expectation can in no wise guide further research" (Geach, 1973, 102). Pietroski and Rey, whose aim it is to save cp laws from the charge of vacuity, call cp clauses "cheques written on the bank of independent theories" (Pietroski and Rey, 1995, 89). If the cheque can be cashed afterwards by giving an independent explanation of the disturbing factor, the cp law has proved non-vacuous.

I wonder whether the cheque is credited to the right account. The better thing to say, to my mind, is that in such cases the *explanation* gets saved from vacuity. The explanation of why the expected effect did not occur may somehow be *prompted* by the cp clause, but since the completing condition is not supplied by any information the clause contains, the connection is quite loose. This looseness is a general weakness of completer accounts. The proviso may stimulate the scientist's imagination, but it is simply not true that it "guides" further research, as Geach demands. The issue under discussion is whether lawful statements qualified with (5) are trivially or non-trivially true, and any rejoinder which changes the wording of the law will change the subject. Hence, completer accounts are not suited to refute the charge of vacuity.[32]

The most exciting reading is arguably (6), the so-called *ceteris absentibus* clause. Geoffrey Joseph, who introduced this term in 1980, gives the following explanation:

> Whether other factors are equal, or constant, is irrelevant. ... What would make laws literally true is not a *ceteris paribus* clause, but rather a *ceteris absentibus* clause. Each of the laws would be true were it restated as: 'Were it the case that all other factors are absent, then, given certain initial conditions, certain resultant conditions would obtain.' (Joseph, 1980, 777)[33]

Joseph eventually rejects *ceteris absentibus* laws, but before considering his reason for doing so, let us take a closer look at the semantics of "*ceteris absentibus*". What distinguishes (6) from (5) is, first, that

[32]Hausman, following Stalnaker's distinction between the meaning of a sentence and its content, which may vary in different contexts, suggests that "*ceteris paribus* clauses have one *meaning* – 'other things being equal,' which in different contexts picks out different *propositions* or *properties*" (Hausman, 1992, 134). His assumption that the context, or rather the speaker's "background understanding" (ibid.), determines what the *cetera* are in each case, has the price of turning the completer account of cp laws into a *pragmatic* view. The charge of triviality, however, is commonly regarded as concerning the semantics and/or the logical form of a law-statement. – Glymour, too, endorses the view that cp claims have a pragmatics which accounts for their (limited) testability (Glymour, 2002).

[33]Johansson (Johansson, 1980, 18) and Hempel (Hempel, 1988, 23 and 29) have taken the same line.

the factors which must be absent can, in a way, be given an independent characterization. If the cetera in question are other *forces*, as both Joseph and Hempel suggest, then we are in possession of what we always looked for: a clause which is "made fully explicit in a finite form" (Earman and Roberts, 1999, 443). Remember the stock objection to the clause "in the absence of other relevant factors", viz. that it "does not assert any determinate relation at all, because it fails to specify which other factors count as relevant" (Lange, 1993, 235). If the excluded factors are other physical forces, this objection is invalidated, for while potential catastrophic interferences cannot be specified in advance, physics does possess a finite and exhaustive list of fundamental physical forces: strong interaction, weak interaction, electromagnetic force, and gravitation.

The second thing to note about the *ceteris absentibus* clause is that it states a counterfactual condition: "*Were* it the case that ...". The idea suggests itself of combining such a clause with the dispositional view of laws of nature – the view that laws say how a system would behave under certain circumstances, e.g. in isolation. We should bear in mind, however, that we were looking for cp clauses as a remedy for the regularity view of laws. It should be obvious that counterfactual clauses cannot play this role, for the simple reason that "a counterfactual uniformity is no uniformity at all" (Cartwright, 1995b, 313).

But, the idea of combining the *ceteris absentibus* clause with the dispositional account of laws deserves to be assessed on its own merits. It faces the following difficulty, brought out by Joseph. As far as we know there is no situation, in the universe we inhabit, where only one of the fundamental physical forces is present. Gravitation, for instance, is omnipresent, there is no shielding from it. Therefore, the *ceteris absentibus* clause is never satisfied, just like clause (2). Supporters of the dispositional view are not worried by this fact, since they hold that laws make hypothetical claims anyway. But the situation is more serious than they think. The *ceteris absentibus* clause does not only establish a counterfactual condition, it posits counter*legal* worlds, i. e., worlds which are nomologically impossible. Joseph explains why:

> [A]ny possible world, distinct from the actual world, that makes the indicative form of a given one of the laws true must be different from every possible world that makes the indicative form of any of the other laws true. This is because the worlds are defined as worlds in which the sole field present is the field mentioned in a given law. There is no possible world in which both F_1 and F_2 are each the only field present. ...In denying the existence of other fields, these worlds deny the existential presuppositions of the remaining laws. (Joseph, 1980, 778)

This is a disastrous result, for the very idea of laws of nature describing counterlegal worlds is an absurdity.[34] Joseph puts it more mildly, expressing his "strong preanalytic intuition that an analysis of the truth conditions for scientific laws must make it possible for all of them to hold in *this* (actual) world" (ibid., 778 f.) To be sure, the problem is not just that the truth conditions of each law depend on those of other laws. Such holism could be true, after all. The deep problem is that the world is not given a chance to make two *ceteris absentibus* laws true at the same time.

Earman and Roberts suggest that the need for provisos "stops at the level of fundamental physics" (Earman and Roberts, 1999, 472). The truth is that it *is* interference all the way down, because it is interaction of forces all the way down. If physicists see no need to add provisos, but leave their laws as they are, this is because they are concerned with force laws, rather than with regularity claims of the kind discussed here. The latter are still plagued with exceptions, since the *ceteris absentibus* clause does not remove disturbing factors by merely stipulating that they be absent. The situation might change if a unified field theory should be developed which reduces the four fundamental physical forces (or three, meanwhile) to one. In an ideally completed physics with a unified force law, the source of disturbances would eventually run dry. But even if this were to happen, it would still be a long way from a unified force law to succession laws describing local regularities.

The clauses (2) through (6) having major drawbacks, I conclude that the first reading, "other things being equal", is the most appropriate one. It captures the linguistic meaning of the Latin phrase, it does not demand the impossible, and it does not make vacuously true the sentences it qualifies. It is unspecified, which is desirable, since cp clauses should be distinguished from the device of specifying antecedents.

[34] Applying the ceteris absentibus clause to the case of interaction between gravitation and electrical charge, Pietroski and Rey ask: "[W]hat counterfactuals shall we consider here? Shall we say that had protons and electrons lacked charge, the force exerted between them would have been equal to Gmm'/d^2? This seems absurd, if even intelligible. ...We doubt that anyone has any idea of what would it be to be a proton or electron without charge, much less how such particles would behave" (Pietroski and Rey, 1995, 105). Joseph makes the same point, using a different example: "Consider a proton in a nucleus of an atom in your finger. What determines its trajectory? Were it not for the strong interaction that overcomes the mutual electromagnetic repulsion between protons, there would be no nucleus. If there were no electromagnetic interaction between protons and electrons, there would be no atom. And if there were no gravitational interaction between atom and earth, the proton would float away along with the rest of you" (Joseph, 1980, 777).

6. Other Things Being Equal – To What?

In discussing some of the literature, I could not avoid speaking of cp *laws.* I would have preferred to speak of cp clauses only. The whole debate about the trivialization charge is based on the assumption that cp clauses are in fact suited to modify lawful statements. I would like to take a step back and challenge this assumption. It is far from obvious that combining a lawful statement with a cp clause yields an intelligible proposition. In the remainder of the paper, I shall frankly advocate the view that, though cp *sentences* have important roles to play, there are no such things as cp *laws.*

So, let us make a fresh start. Philosophers of science have too long been absorbed in asking *which* other things are supposed to be equal. To my mind, the more pressing question is: equal to what? This question is hardly ever raised. Stephen Schiffer addresses the issue when he labels "deceptive" a ceteris paribus sentence such as "If a person wants something, then, all other things being equal, she'll take steps to get it". The sentence, according to Schiffer, "looks as though it's expressing a determinate proposition, because it looks as though 'all other things' is referring to some contextually determinate things and 'equal' is expressing some determinate relation among them" (Schiffer, 1991, 2). It's the last part of the quote which deserves attention. Which "determinate relation" does the word "equal" express in alleged cp laws? Agreed that it's certain conditions, or circumstances, which are supposed to be equal, or unchanged, the question is still: equal to what?

In a singular conditional about a particular situation, the answer would be obvious. Take the counterfactual conditional, "If I had not thrown the stone, the window would not have broken", which invites the objection that this is only true other things being equal. Equal to what? Equal to the factual circumstances of the event described in the antecedent. I did throw the stone, and as the definite article indicates, I did so in a particular situation – in specific, albeit undescribed, circumstances. The stone hit the window and broke it. Still, the truth of the unqualified counterfactual may be questioned. There could have been, say, an earthquake, so that the window would have broken even without the stone. In that very situation, however, there was no earthquake. And this is what the clause "other things being equal" does here: It fixes the circumstances that *actually obtained.* The equality condition has an anchorage, as it were, and therefore the cp clause has a determinate content. The hedged counterfactual reads: "If e had not occurred, *and if everything else had been as it was when e actually occurred,*[35] *f* would not have occurred".

In this counterfactual, "equal" has the meaning of "unchanged". The clause demands that the further circumstances remain as they are in the actual situation. Incidentally, this temporal reading is the only one that *Webster's Dictionary* reports: English usage of "ceteris paribus" expresses the provision that "all other relevant things, factors or elements remain unaltered".[36]

Now it might be objected that the counterfactual I cited is true simpliciter, i.e. without the cp clause. On almost every view of the semantic of counterfactuals, we are taken only to the nearest possible world in which the antecedent holds, and this world does not contain earthquakes or other catastrophic interferences.

I agree that the nearest possible world in which the stone is not thrown does not contain an earthquake, but simply an unbroken window. But why is this so? The plain answer is that we do not have to consider earthquakes because in the situation at hand, none was in the offing. When reasoning counterfactually about particular situations, we do not brood over standards of comparative overall similarity, rather we simply refer to the facts that actually obtained up to the moment described in the antecedent. This direct reference to a singular, actual situation can be brought into the open by using demonstratives: *this* stone, *that* window. This is precisely what Goodman, in his early discussion of counterfactuals, did. One of his examples read: "If that match had been scratched, it would have lighted".[37] It is only because a particular match in particular circumstances is referred to that we accept the counterfactual as true, for in general, scratching is not sufficient for the lighting of matches. But in the case at hand, the match *was* dry enough, the atmosphere *did* contain oxygen, etc. These circumstances do not get described, but they obtained, and the demonstrative reference exploits their determinacy. Though Goodman does not lay emphasis on the fact that his examples contain demonstratives, this fact strikes me as essential. Goodman does not supplement his conditional with a cp clause, but we may say that

---

[35] More precisely "when" means "up to the occurrence of *e*", for the cp clause must leave room for the immediate consequences, causal and logical, of *e*'s nonoccurrence. In other words, the nearest possible world is one which departs from the actual world just at the moment of *e*'s occurrence. I am in agreement with Lewis here: "To get rid of an actual event *e* with the least over-all departure from actuality, it will normally be the best not to diverge at all from the actual course of events until just before the time of *e*" (Lewis, 1986, 171).

[36] *Webster's Ninth New Collegiate Dictionary*, Springfield, Mass. 1983.

[37] The definite articles I used in my example may be looked upon as "degenerate demonstrative singular terms" (Quine, 1960, 102), taking up Russell's observation that "*the*, when it is strictly used, involves uniqueness".

in such conditionals, the demonstrative reference works as a *substitute* for the cp clause, or as an *implicit* cp clause.[38] The hearer understands that in the given situation the requisite conditions obtained, so that explicit conditioning upon certain circumstances becomes redundant, just as Goodman suggests. The demonstrative reference ties the context of evaluation to the context of utterance.

To be sure, cp conditionals need not be in the subjunctive mood. "If I pour this acid into the test-tube, the litmus tincture will turn red, other things being equal." In this indicative, future-directed conditional, the proviso plays the same role as in the counterfactuals above. As long as the antecedent describes a singular situation, the clause can be spelled out as "and if everything else remains unaltered".[39] And if someone accepts the conditional as true even without such a clause, he does so because he understands the constancy assumption as a built-in feature of the conditional.

But we must not lose sight of our original concern. We were out for cp *laws*. Thus, a hedged *universal* statement is called for. The analogous formulation would have to be: "Whenever Fx, then, other things remaining unchanged, Gx". Such conditionals sound familiar, but they invite the hard question: unchanged with respect to what?! No circumstances that could remain unchanged or equal are mentioned in a universally quantified conditional, and none could be mentioned, for making a demonstrative reference to an individual situation is not admissible in a genuinely universal statement. Hence, the cp clause has no anchorage here. It is floating free. If spelled out as a constancy requirement, the cp clause is a foreign element in lawlike sentences, syntactically unfit to qualify universal propositions.

So, under the assumptions that the logical form of a law is a universally quantified conditional, and that "ceteris paribus" is to be translated as "other things being equal", there are no such things as ceteris paribus laws, for reasons of logical form. Cp clauses have singular propositions as

[38] Goodman says: "Notice especially that our assertion of the counterfactual [sc. "If that match had been scratched, it would have lighted"] is not conditioned upon these circumstances obtaining. We do not assert that the counterfactual is true *if* the circumstances obtain; rather, in asserting the counterfactual we commit ourselves to the actual truth of the statements describing the requisite relevant conditions" (Goodman, 1954, 8).

[39] Eventually, the word "else" (or "other"), hitherto mysterious, makes sense. Everything remains unchanged except the change described in the conditional. The objection to clause (2), that all things are never equal since history does not repeat itself, loses its point now. The cases compared are no longer two instantiations of a law, but two singular events or states, one described in the antecedent and one described in the consequent, and the alteration exempted from the constancy requirement is just the change between them. Clause (2) gets vindicated, eventually.

their habitat, not general ones. A cp clause spelled out as a constancy requirement makes a demonstrative reference to particular circumstances, while such circumstances cannot be referred to in law-statements. Constancy requirements have their anchorage outside of law-statements, and it's only because of their indexical nature that they have a determinate content at all.[40]

Why has this fact persistently evaded our notice? One reason could be that some cp conditionals are so closely related to corresponding law-statements. Cp clauses need a reference situation, and when formulating an alleged cp law, there is a natural candidate for that reference situation, viz. the *experimental* situation which gives reason to the formulation of the law, or which confirms it.

The case of experimentation shows quite clearly what the original habitat of cp clauses is, and which role they play there. Bringing about f by doing e worked once, so the experimenter frames a hypothetical law. Experiments are supposed to be repeatable, and in trying to do it again, the scientist must keep constant, or reproduce, the circumstances which obtained the first time. Sometimes he fails, and some of his failings he explains by the conjecture that other things were not equal. In testing quantitative laws, perfect match between the measured result and the predicted result is hard to achieve. The most the scientist can expect is that the gap will progressively decrease the more exactly he reproduces the original circumstances. Galilean idealization has a crucial role to play here. All of this is familiar. The important thing to note is that the phrase "the circumstances which obtained the first time" makes no contact in lawlike sentences.

The upshot is that cp clauses have no business in laws, but only in singular propositions. Although this fact has mostly gone unnoticed, there are a few hints of it in the literature. As quoted above, Schiffer admits cp sentences, but not cp laws. Cartwright comes close to the negative part of my conclusion when she distinguishes "between the descriptions that belong inside a law statement and those that should remain outside as a condition for the regularity described in the law to obtain" (Cartwright, 1995a, 278-9; see also Earman, Roberts and

[40]Still, there seems to be one more use of "other things being equal", where the objection that the equality requirement makes no contact ceases to apply. Consider "Other things being equal, if one boat is newer than another it will be more expensive." Here, the *cetera* are other factors which could affect the prices of boats. This "comparative interpretation" of the cp clause (Morreau, 1999, 171) requires that the two compared items are both mentioned, so that the question "equal to what?" has an obvious answer: equal in both cases. But, no cp *laws* are likely to emerge from such examples. In effect, the sentence doesn't say more than that for boats age is a price-affecting factor.

Smith, 2002, 287). The more specific insight into the *indexical* nature of cp clauses is foreshadowed in a remark of Quine's[41] and in Morreau's "pragmatic" paraphrase of the cp clause.[42] None of them, however, draws conclusions for the notion of a cp law.

Generally, the logical difficulty of combining a general statement with a constancy clause is being overlooked. The capricious way in which the cp clause is inserted into law-statements by various authors is telling. Every conceivable placement can be found: Kincaid, Pietroski and Rey let the law-statement begin with the cp clause, Kurtzman writes the clause between the quantifier and the antecedent, Silverberg and Schiffer behind the sentential connective, Fodor sometimes before the law, sometimes at the end, Cartwright mostly into the antecedent, and for Schurz the clause is part of the connective, which he takes as a "normic" conditional operator. One should think that it makes a difference whether a constituent of a law-statement is part of the antecedent or of the consequent, or whether it is quantified over or not, or whether it is itself a quantifier. Once we acknowledge that the cp clause is not a logical constituent of the law, the diversity of positionings becomes less amazing. On most of the readings discussed above, the cp clause has the logical form of an additional if-clause. Syntactically, such a clause may in principle be placed before, within or behind the law-statement, the effect always being that it sets an additional condition that limits the number of cases covered by the law. But on closer examination, the clause can only play that role if the constancy condition has an anchorage somewhere: equal to what? The only intelligible answer I found goes: "equal to the factual circumstances of the event mentioned in the antecedent". This answer works only with singular conditionals, not with general ones. In singular conditionals, the clause has a determinate content, insofar as it taps the determinacy of the situation which actually obtained. It is worth noting that such singular conditionals play an important role in counterfactual analyses of event causation, though their singularist character has not always been recognized.

We arrive at counterfactuals of the sort "If e had not happened, and if everything else had been as it was when e actually happened, f would not have happened". These conditionals are perfectly intelligible, but they

[41]Quine observes that the "clues to the scope of *'ceteris paribus'* ... are afforded by the context or other special circumstances of the particular utterance" (Quine, 1960, 225).

[42]Cp clauses, says Morreau, "can be used to hedge claims against the possibility of changed circumstances" (Morreau, 1999, 165). His "pragmatic" paraphrase "ties the interpretation of the modifier to some context of evaluation; for a thing, factor or element to 'remain unaltered', it must remain as it is *there*" (ibid., 166). An observation of Hausman's (see above, fn. 32) also points into that direction.

cannot be turned into laws. A law cannot be combined with an indexical constancy clause, since the clause has its required relatum outside the law, and the resulting hybrid statement, half-singular and half-universal, would be hard to make sense of. Given the assumptions I made, there are no such things as cp laws.

By the same token, cp clauses cannot be used to restore the truth of succession laws in the face of exceptions. Thus, they cannot be used to support the cause-law thesis. But as I just pointed out, they can do something better for the theory of causality. Cp clauses retain their function when combined with singular counterfactuals, and they actually do a good job in the counterfactual theory of event causation.[43] The truth conditions of causal counterfactuals cannot be given, of course, by corresponding laws, since strict causal laws are nowhere to be found, while true causal counterfactuals abound. Not being supported by strict laws of succession, causal counterfactuals have to stand on their own feet. But that's a different matter.

It might be objected that this sketch of a solution works only for the first of the six readings of "*ceteris paribus*" which I distinguished. This is true, but I must remind the reader that I had rejected the other paraphrases for independent reasons.

Moreover, I had narrowed down my attention to empirical laws of succession. So in a way, I agree with the view that "*ceteris paribus* stops at the level of fundamental physics" (Earman and Roberts, 1999, 472). Exceptions being counterinstances, the notion of exception has no application to laws which do not purport to describe what actually happens. Interpreted in a certain way, force laws need no hedging. They can remain true even if in the real world forces always interact.

Earman, Roberts and Smith have argued that since cp claims are "open-ended" and have no determinate content, they cannot be laws. I agree with their conclusion that there are no cp laws, but not with their view of cp clauses. In the counterfactuals at hand, cp clauses are not open-ended. They do have a determinate content and determinate truth conditions. But since this content is indexical, cp clauses are unfit for being combined with laws. The linguistic meaning of "ceteris paribus" is "other things being equal", while the clause gets its context-dependent propositional content by exploiting a particular context of utterance.

[43] I elaborated on a counterfactual account of event causation which appeals to singular counterfactuals with indexical cp clauses in my (Keil, 2000, 261–279 and 431-457).

References

Armstrong, D. (1983). *What is a Law of Nature?* Cambridge.

Bhaskar, R. (1978). *A Realist Theory of Science.* Hassocks, Sussex, 2 edition.

Cartwright, N. (1983). *How the Law of Physics Lie.* Oxford.

Cartwright, N. (1989). *Nature's Capacities and their Measurement.* Oxford.

Cartwright, N. (1992). Aristotelian natures and the modern experimental method. In Earman, J., editor, *Inference, Explanation and Other Philosophical Frustrations*, pages 44–71, Berkeley.

Cartwright, N. (1995a). Ceteris paribus laws and socio-economic machines. *The Monist*, 78:276–294.

Cartwright, N. (1995b). How laws relate what happens: Against a regularity account. In Stachowiak, H., editor, *Pragmatik. Handbuch pragmatischen Denkens*, volume V, pages 305–314, Hamburg.

Cartwright, N. (1999). *The Dappled World. A Study of the Boundaries of Science.* Cambridge.

Cummins, R. (1983). *The Nature of Psychological Explanation.* Cambridge, Mass.

Davidson, D. (1980a). *Essays on Actions and Events*, chapter Mental Events, pages 207–225. Oxford.

Davidson, D. (1980b). *Essays on Actions and Events*, chapter Actions, Reasons and Causes, pages 3–19. Oxford.

Davidson, D. (1993). Replies. In Stoecker, R., editor, *Reflecting Davidson*, Berlin/New York.

Davidson, D. (1995). Laws and cause. *Dialectica*, 49:263–279.

Dretske, F. (1977). Laws of nature. *Philosophy of Science*, 44:248–268.

Ducasse, C. J. (1968). On the analysis of causality [1957]. In his *Truth, Knowledge and Causation*, pages 15–20. London/New York.

Earman, J. and Roberts, J. (1999). *Ceteris paribus*, there is no problem of provisos. *Synthese*, 118:439–478.

Earman, J., Glymour, C. and Mitchell, S. (2002). Editorial. *Erkenntnis*, 57:277–280.

Earman, J., Roberts, J. and Smith, S. (2002). *Ceteris paribus* lost. *Erkenntnis*, 57:281–301.

Fodor, J. A. (1991). You can fool some of the people all of the time, everything else being equal. Hedged laws and psychological explanations. *Mind*, 100:19–33.

Geach, P. T. (1973). Aquinas. In: G. E. M. Anscombe and P. T. Geach *Three Philosophers.* Oxford, pages 64–126.

Giere, R. N. (1988). Laws, theories, and generalizations. In Grünbaum, A. and Salmon, W., editors, *The Limitations of Deductivism*, pages 37–46, Berkeley/London.

Glymour, C. (2002). A semantics and methodology for ceteris paribus hypotheses. *Erkenntnis*, 57:395–405.

Goodman, N. (1954). *Fact, Fiction, and Forecast.* London.

Guarini, M. (2000). Horgan and Tienson on ceteris paribus laws. *Philosophy of Science*, 67:301–315.

Hausman, D. M. (1992). *The Inexact and Separate Science of Economics.* Cambridge.

Hempel, C. G. (1988). Provisos: A problem concerning the inferential function of scientific theories. In Grünbaum, A. and Salmon, W. C., editors, *The Limitations of Deductivism*, pp. 19–36, Berkeley/London.

Horgan, T. and Tienson, J. (1996). *Connectionism and the Philosophy of Psychology.* Cambridge, Mass./London.

Hüttemann, A. (1998). Laws and dispositions. *Philosophy of Science*, 65:121–135.

Johansson, I. (1980). Ceteris paribus clauses, closure clauses and falsifiability. *Journal for General Philosophy of Science*, 11:16–22.

Joseph, G. (1980). The many sciences and the one world. *Journal of Philosophy*, 77:773–791.

Keil, G. (2000). *Handeln und Verursachen.* Frankfurt am Main.

Kincaid, H. (1996). *Philosophical Foundations of the Social Sciences.* Cambridge.

Klee, R. (1992). Anomalous monism, ceteris paribus and psychological explanation. *British Journal for the Philosophy of Science*, 43:389–403.

Kurtzman, D. R. (1973). Ceteris paribus clauses: Their illumination and elimination. *American Philosophical Quarterly*, 10:35–42.

Lakatos, I. (1978). *Philosophical Papers*, volume I, chapter Fallibilism and the Methodology of Scientific Research Programmes, pages 8–101. Cambridge.

Lange, M. (1993). Natural laws and the problem of provisos. *Erkenntnis*, 38:133–148.

Lewis, D. (1986). *Philosophical Papers.* Oxford.

Lipton, P. (1999). All else being equal. *Philosophy*, 74:155–168.

Mackie, J. L. (1980). *The Cement of the Universe. A Study of Causation* [1974]. Oxford, 2 edition.

Mill, J. S. (1973). *A System of Logic, Ratiocinative and Inductive* [1843]. Toronto, 4 edition. (Collected Works, Vol. VII/VIII).

Morreau, M. (1999). Other things being equal. *Philosophical Studies*, 96:163–182.

Nagel, E. (1961). *The Structure of Science.* London.

Persky, J. (1990). Ceteris paribus. *Journal of Economic Perspectives*, 4:187–193.

Pietroski, P. and Rey, G. (1995). When other thing's aren't equal: Saving ceteris paribus laws from vacuity. *British Journal for the Philosophy of Science*, 46:81–110.

Popper, K. R. (1974). Replies to my critics. In Schilpp, P. A., editor, *The Philosophy of Karl Popper, Bk. II*, pages 961–1197, La Salle, Ill.

Quine, W. V. O. (1960). *Word and Object.* Cambridge, Mass.

Robbins, L. (1984). The nature and significance of economic science [1935]. In Hausman, D. M., editor, *The Philosophy of Economics*, pages 83–110, Cambridge.

Russell, B. (1948). *Human Knowledge. Its Scope and Limits.* New York.

Russell, B. (1986). On the notion of cause [1912]. In his *Mysticism and Logic*, London.

Schiffer, S. (1991). Ceteris paribus laws. *Mind*, 100:1–17.

Schurz, G. (2001a). Pietroski and Rey on ceteris paribus laws. *British Journal for the Philosophy of Science*, 52:359–370.

Schurz, G. (2001b). What is 'normal'? An evolution-theoretic foundation of normic laws and their relation to statistical normality. *Philosophy of Science*, 68:476–497.

Scriven, M. (1959). Truisms as the grounds for historical explanations. In Gardiner, P., editor, *Theories of History*, pages 443–475, New York/ London.

Scriven, M. (1961). The key property of physical laws – inaccuracy. In Feigl, H. and Maxwell, G., editors, *Current Issues in the Philosophy of Science*, pages 91–101, New York.

Silverberg, A. (1996). Psychological laws and non-monotonic logic. *Erkenntnis*, 44:199–224.

Smith, S. (2002). Violated laws, ceteris paribus clauses, and capacities. *Synthese*, 130:235–264.

Steiner, M. (1986). Events and causality. *Journal of Philosophy*, 83:249–264.

Swartz, N. (1985). *The Concept of Physical Law.* Cambridge.

Tye, M. (1992). Naturalism and the mental. *Mind*, 101:421–441.

van Fraassen, B. C. (1989). *Laws and Symmetry.* Oxford.

Warfield, T. (1993). Folk psychological ceteris paribus laws? *Philosophical Studies*, 71:99–112.

Woodward, J. (2000). Explanation and invariance in the special sciences. *British Journal for the Philosophy of Science*, 51:197–254.

NECESSARY LAWS

Max Kistler

Within the empiricist tradition, it is often taken for granted that the laws of nature are contingent. According to this view, metals could contract upon heating instead of expanding as they actually do. I shall attack this view first, by questioning an essential assumption on which it depends, and second by giving a positive reason to think that at least some laws are necessary. I begin by looking a little closer at the reasons for the contingency view of laws.

1. Lewis' and Armstrong's combinatorialism and the contingency theory of laws

David Lewis and David Armstrong adopt radically opposite positions with respect to the metaphysical interpretation of modality. For Lewis, there are other possible worlds that are just as real as ours. For Armstrong, other possible worlds are ways to combine the particulars and universals of our actual world into states of affairs. For Armstrong, there is only one actual world, which is the one we inhabit. It is absolutely actual. For Lewis, actuality is a relative notion: our own world is indeed actual, but only relatively to us. Actuality is an indexical, context-sensitive concept that picks out different worlds at different world-contexts. In this world Lewis is a philosopher, but some other world represents him as being a plumber. The world inhabited by Lewis the plumber is the actual world for Lewis the plumber although it is only a possible world for Lewis the philosopher[1]. From the viewpoint of Lewis' realism with respect to possible worlds, Armstrong is a "linguistic ersatzer" who holds that the conceptual work done by postulating possible worlds can be done by linguistic constructions. If it can be done - a thesis Lewis (Lewis, 1986a) denies – parsimony dictates to prefer a

[1]Lewis denies overlap with respect to individuals between possible worlds, so Lewis the plumber is not the very same individual as Lewis the philosopher but only his counterpart in virtue of some appropriate similarity relation. See below, section 4.

J. Faye, P. Needham, U. Scheffler and M. Urchs, (eds.), Nature's Principles, 201–227.

sparse metaphysical doctrine with only one real world to a luxurious doctrine with an infinity of real worlds.

Armstrong and Lewis also adopt radically different positions with respect to the issue of the metaphysical nature of laws. Here Armstrong is the realist who thinks of laws as objectively existing relations between universals[2] whereas Lewis adopts the Humean or anti-realist position that there is no necessity in nature. Laws are defined relatively to a hypothetical ideal science: the laws are the axioms and theorems of an ideal scientific theory[3].

In spite of this important disagreement in the metaphysical interpretation of both possible worlds and laws, Armstrong and Lewis agree in following the pretheoretical intuition that the laws of nature are contingent. Even if in our actual world it is a law that all metals expand when their temperature rises, both hold that it is, nevertheless, possible that a metallic object contracts upon heating. Instead of relying directly on intuitions, both Lewis and Armstrong argue for the thesis of the contingency of laws with some version of the principle of combination. There is however an important difference between Lewis' "principle of recombination" (Lewis, 1986a, p. 87) and Armstrong's (Armstrong, 1989) combinatorialist theory of possibility. For Armstrong, possibilities arise from combining logical atoms, which are the constituents of states of affairs, i.e. particulars and universals. For Lewis, possibilities arise from the rearrangement of the distribution of intrinsic qualities over space-time locations. As we shall see, this difference has important consequences for the issue of the modal status of laws.

At first sight, David Lewis' position with respect to the issue of the modal status of laws seems straightforward: "There might have been altogether different laws of nature" (Lewis, 1986a, p. 1); "there are [...] worlds where [...] totally different laws govern the doings of alien particles with alien properties" (Lewis, 1986a, p. 2). Speaking of the rival doctrine of "strong laws" according to which laws are necessary, he says: "If a theory of strong laws is to be credible, it had better provide not only a sense of 'possible' in which violations of laws are impossible, but also another sense in which violations of laws are possible. Perhaps that second sense cannot be provided. In that case the doctrine of strong laws is not credible enough to deserve consideration." (Lewis and Langton, 1998, p.122) Lewis's reason for holding that the laws are contingent lies

[2]Cf. Armstrong (Armstrong, 1983). Dretske (Dretske, 1977) and Tooley (Tooley, 1987) hold similar positions.

[3]This is the famous "best-system analysis of laws" which Lewis adopts from Ramsey. Cf. Lewis (Lewis, 1973; Lewis, 1983; Lewis, 1994).

in his adherence to the fundamental doctrine of Humean supervenience according to which "all there is to the world is a vast mosaic of local matters of particular fact, just one little thing and then another. [...] For short: we have an arrangement of qualities. And that is all. There is no difference without difference in the arrangement of qualities. All else supervenes on that." (Lewis, 1986b, pp. ix f.) This doctrine implies that objectively, laws are nothing but regularities. To distinguish them from accidental coincidences, the best-system analysis says that, among all regularities, the laws are those that science will eventually pick out as axioms and theorems of the ideal theory. It is an essential part of this Humean doctrine that there are no necessary connections between what happens at different points, between the qualities instantiated at different spatio-temporal locations. A quality instantiated at one point imposes no modal constraint whatsoever on the qualities instantiated elsewhere. Anything can possibly be juxtaposed to anything. To take Lewis' example, it is a lawful regularity obtaining in this world that bread-eating prevents starving (Lewis, 1986a, p. 91). But this regularity might possibly not obtain. There are worlds in which I eat bread and nevertheless starve. Lewis generalizes this idea into a "principle of recombination [...] Roughly speaking, the principle is that anything can coexist with anything else, at least provided they occupy distinct spatiotemporal positions. Likewise, anything can fail to coexist with anything else." (Lewis, 1986a, pp. 87f.)

One important difference with respect to Armstrong's version of the doctrine of the contingency of laws is the following. Lewis' thesis of the contingency of laws applies directly only to laws implying necessary connections between qualities instantiated at *different* space-time locations. The "distinct existences" (to use Armstrong's terminology) that can be, according to Lewis, combined in all ways are qualities that fully occupy a location. This implies that Lewis' Humean combinatorialism does not automatically classify *all* laws as contingent. The thesis that qualities instantiated at different spatio-temporal locations are only contingently linked, implies that what are often called *causal* laws are contingent. Such laws link what happens at different spatio-temporal locations. This is the result Lewis focuses on when he says that the principle of recombination settles "the question whether laws of nature are strictly necessary. They are not; or at least laws that constrain what can coexist in different positions are not" (Lewis, 1986a, p. 91).

However, not all laws are causal laws, and Lewis' Humean principle of recombination of qualities instantiated at different locations does not imply that these other laws, and in particular, the laws of association are contingent. Lewis notes only in passing, and somewhat tentatively, that his strategy "to take a Humean view about laws and causation, and use it instead as a thesis about possibility" (*ibid.*) does not imply free combinability with respect to the qualities that are co-exemplified at the same space-time location. Thus, Lewis admits that "perhaps" (*ibid.*) his Humean argument entails the contingency of laws only "with the exception of laws constraining what can coexist at a single position, for instance the law (if such it be) that nothing is both positive and negative in charge." (*ibid.*) In fact, having noted that the Humean doctrine doesn't force upon him the contingency of such laws, Lewis says that we have no means to know whether such incompatibilities are necessary or not. This is a question on which

> there seems to be no way at all of fixing our modal opinions, and we just have to confess our irremediable ignorance. I think one question of this kind concerns incompatibility of natural properties. Is it absolutely impossible for one particle to be both positively and negatively charged? Or are the two properties exclusive only under the contingent laws of nature that actually obtain? I do not see how we can make up our minds; or what guarantee we have that there must be some way to settle the question. [...] Whatever the truth may be, it isn't up to us. (Lewis, 1986a, p. 114)[4]

His agnosticism about (epistemically) possible natural incompatibilities makes Lewis combinatorialism weaker than Armstrong's. For Lewis, it may be the case that the quality of being positive in charge is not a combinatorial atom. If it is not, then it cannot combine with the quality of being negative in charge at the very same time and place. In that case there would be a link of natural necessity between different qualities instantiated at the same time and place. For Armstrong (Armstrong, 1989), such necessary relations can only have a logical or mereological origin[5]. We can express the difference in the following way: Armstrong takes possibilities to arise from combinations of "distinct existences" which are bare particulars and simple universals whereas Lewis takes possibilities to arise from combinations of "distinct existences" which are "thick particulars": the totality of all qualities instantiated at a given space-time point. Within one combinatorial atom, there may be links of necessity. It follows from this difference in their respective accounts of independence and combinatorial possibility that the existence

[4]I disagree. As I shall argue below, our ignorance in this case is not irremediable. It stems from our scientific ignorance of whether these properties are two different determinates of a given determinable. If they are then it is logically and thus metaphysically impossible that they are instantiated at the same time and place.

[5]I shall criticize this thesis of Armstrong's below, in section 6.

of necessary laws is compatible with Lewis' metaphysical framework, but not with Armstrong's.

2. Do universals have a quiddity, suchness or haecceity?

One can understand the debate on the modal status of laws in two ways. On a strong reading of the thesis of the necessity of laws, they hold in *all* possible worlds. I agree with Armstrong and Lewis that this is not plausible. The controversial thesis that I try to defend against Armstrong and Lewis is weaker. It says that all possible worlds that share our universals also share our laws. The laws are necessary in the weak sense that they hold in all those possible worlds that share our actual universals. Armstrong and Lewis (and many others, such as Mellor, 1993 and Mellor, 1997) however think that laws are not even necessary in this weaker sense, and that there are possible worlds that share our universals but not our laws. According to this doctrine of the contingency of laws, a given universal F that is nomically linked to a universal G in the actual world is not linked to G in some other possible world. Although it is a law in the actual world that metals expand when heated, there are possible worlds where it is instead a law that metals contract when heated. The contingency thesis I am arguing against holds not only that some possible worlds do not share our laws (this is a thesis I accept and which corresponds to the denial of the strong thesis according to which our laws hold in *all* possible worlds), but that the very same property of being metallic which exists in the actual world might be differently related to other properties, such as expanding and contracting, than it is in the actual world.

I shall now offer the following argument against the contingency view of laws. Presupposing for the sake of this discussion that laws are relations between universals, the contingency theory holds that one universal might have different nomical relations to other universals than it has in the actual world. If laws are contingent, universals are embedded in different laws in different possible worlds. Consider an actual universal U and a non-actual universal U^* in some heteronomic world. According to the contingency view, U and U^* may be identical although their nomological properties differ, in other words, although U and U^* are nomically linked to different properties. There seem to be two ways to construe such a cross-world identity claim for universals. According to the first, one universal can literally exist in more than one possible world. In section 4, I shall argue that this assumption leads to the conclusion that at least some laws are necessary.

According to the second way to construe the trans-world identity of U and U^*, they are the same universal although they are not logically and numerically identical (in so far as they differ with respect to the laws in which they are embedded) because they have the same essence[6]. In this section and the following, I attack two arguments for the existence of such an essence. According to this construal of the contingency view, U and U^* are different ways the same universal could have been: they differ nomologically but share a non-qualitative[7] essence, something which has been variously called "haecceity" (Rosenkrantz, 1993), "suchness" (O'Leary-Hawthorne and Cover, 1997) or "quiddity" (Armstrong, 1989). Laws can then be considered as contingent relations between universals because the identity of the universals is independent of the laws, being determined instead by their quiddity. How can we make sense of the hypothesis that universals have a quiddity or non-qualitative essence? One way is to conceive of the quiddity of a universal in a purely formal way. It can equally well be applied to argue for the "haecceity" of individuals, and indeed for the haecceity of anything at all that exists. Each individual, says Rosenkrantz (Rosenkrantz, 1993), has its own haecceity, in virtue of the simple fact of being identical with itself. Given that it is true for every x that $x = x$, one can consider "$= x$" as equivalent to the predicate "being identical to x". Then $(\forall x)(x = x)$ is equivalent to

$$(\forall x) \quad F_x x$$

where "F_x" is the predicate "is identical to x". Rosenkrantz takes it for granted that one is ontologically committed to the reference of the predicate, i.e. to the existence of the *property* F_x which is x's haecceity. In other words, he proves the existence of a haecceity F_x for each and every individual x, by supposing that every predicate expresses a property, at least if the predicate is satisfied by something. Given that for each x there is something satisfying the predicate F_x, there is a property

---

[6]Mere similarity is not sufficient for identity. Without an essence, with all properties equally contributing to the identity of a universal, only perfect similarity, i.e. having all properties in common, is sufficient for identity. This is equivalent to the first option.

[7]There seems to be still another possibility. The essence of a universal could consist of part of its properties. As I shall argue below, all second-order properties of a universal are nomological properties, i.e. nomic links to other universals. Therefore, such an essence would consist of part of the nomological properties of the universal. I think this option must be ruled out because there is no principled reason why some of the laws in which a universal takes part, should be more essential to its identity than others.

expressed by that predicate which is x's haecceity[8]. As a definition of what it is to be the haecceity of an individual, Rosenkrantz offers:

> F is a haecceity $=_{df.}$ $(\exists x)(F$ is the property of being identical with $x.)$ (Rosenkrantz, 1993, p. 3)

With respect to universals, e.g. Redness, he argues in an analogous, purely formal way, that the proposition "that $(\exists x)(x$ is red$)$" (Rosenkrantz, 1993, p. 12) implies the proposition "that $(\exists x)$ $(x =$ Redness$)$" (*ibid.*).

This inference presupposes the thesis, already implicitly relied on in the argument for the haecceity of individuals, that any use of a predicate carries ontological commitment to the reference of the predicate. To arrive at the existence of the haecceity of the universal Redness, Rosenkrantz uses the same argument once again, but on a higher ontological level[9]. He argues that in

$$(\exists x)(x = \text{ Redness })$$

one can consider "=Redness" as a predicate F_R, and then spell out the presupposed ontological commitment to the property F_R refers to. This is the haecceity (or quiddity) of Redness, "the property of being identical to Redness".

I think one can grant that there is a sense in which such properties exist. They belong to what Lewis calls "abundant" (Lewis, 1986a, p. 59; Lewis, 1983, pp. 345f.) properties and opposes to the "sparse" or "natural" properties. Natural properties are such that it can only be found out a posteriori that they are exemplified. The property of being

---

[8] In the case of particulars, Rosenkrantz (Rosenkrantz, 1993, chap. 2), following Adams (Adams, 1979), adds a less formal argument for the existence of haecceities, from the possibility of a world containing strictly indiscernible individuals: if they are nevertheless numerically different, postulating a haecceity for each is the only possible explanation available. I have two objections against this argument: First, such an argument cannot establish that haecceities *actually* exist; second, it is a non sequitur even with respect to those possible worlds where there are strictly indiscernible individuals. As the reasoning about quantum mechanical indistinguishable particles below (section 3) shows, such a possibility only shows that countability does not always go together with individual identity; such individuals are numerically more than one but this alone does not suffice to establish that necessarily each has its own individual identity. Cf. (Lowe, 1998).

[9] In the case of universals, Rosenkrantz (Rosenkrantz, 1993, p. 132; and note 55) explicitly says that such a formal argument suffices if it combined with the principle according to which "necessarily, if something has a haecceity, then everything has a haecceity" (Rosenkrantz, 1993, p. 13). I think that universals do not obey the same criteria of identity as particulars. Therefore, it is not obvious that arguments in favour or against haecceity carry over from the case of particulars to the case of universals. A crucial relevant difference is that it makes sense to say that there are (e.g. in a quantum-mechanical system) countably many indistinguishable particulars, but not to say that there are countably many indistinguishable universals. See below, section 3.

red is a natural property in this sense. However, once the existence of the property of being red is granted, it is a matter of pure logic to show that redness has a haecceity in Rosenkrantz' sense, by showing that it is identical with itself. A haecceity, so understood, is not a natural property, for insofar as some entity x exists, logic alone suffices to establish that x has a haecceity.

However, Rosenkrantz' shadowy haecceities cannot ground transworld identity of universals embedded in different laws, and therefore cannot help justify the contingency view of laws. For on Rosenkrantz' construal of haecceities, they are not associated with criteria allowing to judge whether two singular expressions referring to haecceities refer to the same or different haecceities. So how could the defender of the contingency of laws rely on them to ground the claim that universals U and U^*, being in different worlds and embedded in different laws, share *the same* haecceity? Rosenkrantz' way to introduce them only guarantees that each has *a* haecceity (in a sense in which any existing entity whatsoever has a haecceity), but not that both have the same haecceity.

3. An argument for quiddity from the possibility of indistinguishable universals

O'Leary-Hawthorne and Cover (O'Leary-Hawthorne and Cover, 1997) give a less formal argument for the thesis that it is at least possible that universals have quiddity (they call it "suchness", leaving "haecceity" for individuals). They ask us to conceive of the possible situation in which there are two indistinguishable universals, Fness and Gness. It may indeed seem plausible at first sight that if it were possible that there exist two numerically different but qualitatively perfectly indistinguishable universals, this would give us a reason to postulate that each has its own nonqualitative quiddity which makes it different from the other.

Against this reasoning, I offer two arguments. First, the hypothesis of two indistinguishable universals violates the Causal Criterion of Identity that follows from another traditional and widely shared metaphysical principle, the Causal Criterion of Reality (CCR)[10]. This latter principle says that something is real if and only if it is capable of making a difference to causal interactions or causal processes[11]. Now the

[10] I develop this argument in Kistler, 2002.

[11] The status of this principle as an ultimate criterion of reality can and has been doubted. Armstrong has argued that it is "not [...] a necessary truth, but merely good methodology" (Armstrong, 1984, p. 256). This is correct but I think that, in metaphysics, good methodology is the only accessible criterion of truth: God's point of view being inaccessible, we are condemned to adopt a naturalistic standpoint. When we make the metaphysical claim that

Causal Criterion of Identity (CCI) follows if the CCR is applied to the properties of an entity. The identity of an entity is determined by those of its properties whose exemplification makes a causal difference. The situation imagined by O'Leary-Hawthorne and Cover's is incompatible with the CCI: For the universals *F*ness and *G*ness to be different, it must be possible that it makes a difference, whether it is *F*ness or *G*ness that is exemplified in a given situation. And if it makes a difference, one has a nomological property the other lacks. From the perspective of the CCI, to say, "there is no guarantee that two universals at a world have different causal powers" (O'Leary-Hawthorne and Cover, 1997, p. 107), just means: There is no guarantee against counting the same universal twice over; but counting it twice does not make it into two really different universals. The CCI gives us grounds to judge them identical if they share all their causal powers. In this clash of doctrines, it seems to me that the burden of proof lies on the opponent of the CCI who postulates that there may be real differences which make no causal difference, because the CCI gains some a priori plausibility from the fact that it is the metaphysical generalisation of a methodological principle, grounding existence claims in science.

My second argument is more significant because less question begging than the first. Even if we granted the possibility that there be two or more indistinguishable yet numerically different universals, this would not justify the conclusion that each has its own quiddity, making them intrinsically different although they are qualitatively indistinguishable. The argument from numerical difference with qualitative indiscernability to a non-qualitative essence (or quiddity or haecceity) is in general not valid, because it can be shown that it is not valid in the case of particulars. So let us consider the analogous case for particulars.

Quantum mechanics teaches that there are systems of interacting fundamental particles of the same type and in the same state that contain a number of perfectly indistinguishable particles. The particles constitutive of such a system are numerically different yet qualitatively indistinguishable. With respect to these particles, let us construct an argument that runs parallel to O'Leary-Hawthorne and Cover's argument for the quiddity of universals. Its premise says that there is a set of partic-

---

x exists, our only justification is that the best available interpretation of all available facts – an interpretation that will be either scientific or compatible with science – gives us reason to believe in the existence of x. Such an inference to the best explanation does not of course establish that those entities, in whose existence we have for the moment no good reason to believe, do really not exist. It is only that there is no rational justification to believe in the existence of anything that neither belongs to the empirical facts nor must be postulated to account for these facts.

ulars that are perfectly indistinguishable. (This corresponds to their premise that there are the perfectly indistinguishable universals *F*ness and *G*ness.) The premise is true for such particles. Nevertheless the following reasoning shows that its conclusion is false: indistinguishable particles do not have any individual haecceity making them intrinsically though non-qualitatively different from each other. Redhead and Teller (Redhead and Teller, 1992) call the hypothetical non-qualitative essence of particulars – what I have called "haecceity" – "Transcendental Individuality" and characterise it as "that unifying principle of an individual which is thought to transcend its attributes" (Redhead and Teller, 1992, pp. 202f.). It "is not another property, but [...] that by which an entity allegedly acquires its identity" (*ibid.*). One of the roles of this concept of a principle of individuation transcending the attributes (or qualitative properties) of a particular – in our words, of a haecceity – is to be "that in virtue of which the individual can bear a label, and that in virtue of which the individual can be thought of as persisting through time as one individual" (Redhead and Teller, 1992, p. 203). However, the quantum statistics that correctly describes such systems of indiscernible particles seems to be incompatible with the hypothesis that the particles have a haecceity and can therefore be labelled differently. The point can relatively simply be brought out in the following case. Consider two indistinguishable particles (more precisely, bosons) each of which can be in one of two pure quantum states $|a^h>$ and $|a^t>$ (as "head" and "tail"). Then if these particles have haecceity, and can thus be labelled 1 and 2, there seem to be four possibilities which should be equally probable:

1 $|a^h_1> |a^h_2>$: both particles are in state $|a^h>$

2 $|a^t_1> |a^t_2>$: both particles are in state $|a^t>$

3 $|a^h_1> |a^t_2>$: particle 1 is in state $|a^h>$ and particle 2 is in state $|a^t>$

4 $|a^t_1> |a^h_2>$: particle 1 is in state $|a^t>$ and particle 2 is in state $|a^h>$

However, this a priori conception of the possible states of the system is incompatible with the best available interpretation of the observed phenomena. Quantum statistics requires counting what appears in this presentation as two possibilities 3 and 4 as in fact only one possibility. This makes an empirical difference for if the classical statistics were true of the system, it would be twice as probable to find one particle in each state, i.e. to find the state "3 or 4" than to find both particles in the same state, $|a^h>$ or $|a^t>$, whereas it is, as a matter of empirical fact, equiprobable. Quantum mechanics accounts for this fact by postulating that our two-particle system does not have four possible states available, as it would have in a classical representation, but only three. These correspond to the states 1 and 2, whereas the third is a mixed state which can be represented as a superposition of the classical states 3 and 4[12].

O'Leary-Hawthorne and Cover (O'Leary-Hawthorne and Cover, 1997) and others take the existence of systems of indistinguishable particles as an argument *in favour* of haecceity[13]. However, Redhead and Teller (Redhead and Teller, 1992) show that it doesn't constitute such an argument, and that on the contrary, the supposition of the existence of haecceities implicit in the labelling of the particles creates a puzzle, which is dissolved by dropping that supposition. The puzzle is that, as long as one rests with the classical representation and its idea that each particle has its individual identity, which justifies attributing a label to it, the states 3 and 4 seem to be genuine possibilities, which are never actualised. However, the existence of such possibilities has no scientific grounding, the appearance of their existence flowing from a metaphysical prejudice in favour of haecceities, reflected in labelling. The empirical fact that such systems obey a statistics that corresponds to the existence of three states suggests that there really are only three possibilities, which can be explained by the hypothesis that such particles do not have any haecceity[14]. The situation seems to plead for

[12]What I have said is true only for bosons, which require a symmetric state description. Fermions whose state description must be asymmetric cannot be in states 1 and 2, but must be in a mixed asymmetric state. Cf. French and Redhead, 1988.

[13]Black presents a famous a priori argument for the same conclusion: It "is logically possible that the universe should have contained nothing but two exactly similar spheres" (Black, 1952, p. 156). Similarly, Adams (Adams, 1979, p. 22) and Armstrong (Armstrong, 1997, p. 108) argue for haecceity in some possible world in which there exist two indistinguishable counterparts of Earth one of which, at a certain time, ceases to exist. As Swinburne (Swinburne, 1995, p. 394) notes, an important weakness of such arguments is that they could at best show that the objects existing in some possible world very distant from the actual one have a haecceity – or "thisness", as Swinburne calls it –, whereas it does not address the question that primarily interests us, whether the material objects in the actual world have a haecceity. Although it fails, the argument from the existence of indistinguishable particles discussed in the text is relevant for this latter question, for quantum mechanics tells us that they exist in the actual world.

[14]This reasoning seems to contradict Lewis' view that "we do not find out by observation what possibilities there are" (Lewis, 1986a, p. 112). If I am correct, then science is more directly relevant for metaphysics than Lewis allows. Even if possibilities are established by a purely a priori logical principle of recombination (Lewis, 1986a, p. 87), it is science that determines the nature of the entities, particulars and properties, to be recombined. The conflict can be resolved by observing that the relevance of science reaches only over the range of nomologically possible worlds whereas Lewis notes the irrelevance of science with respect to the determination of the full range of possibility, nomologically impossible possibility included.

Lowe's (Lowe, 1998, p. 193) thesis that identity and countability do not always go together: two indistinguishable particles do not have their own individual identity although they are countable as two.

If this reasoning is correct, quantum mechanics shows that numerical difference does not suffice to establish the existence of a non-qualitative essence, or haecceity, in the case of particulars. Therefore this is not a valid argument pattern that can be used, as O'Leary-Hawthorne and Cover (O'Leary-Hawthorne and Cover, 1997) do, to argue in the case of universals that the hypothesis of two numerically different yet indistinguishable universals would imply that such universals would have quiddity. To conclude the reasoning of the two preceding sections, we have seen that O'Leary-Hawthorne and Cover's argument for the existence of a non-qualitative essence of universals is invalid, whereas the kind of essence of universals whose existence Rosenkrantz' argument allows to establish, is too weak to be able to ground the contingency view. For we have seen that the contingency view needs the essence of universals as a ground for their identity across different possible worlds, and Rosenkrantz' haecceities are not up to that task. Let us now look at the second and stronger way to construe the identity of universals across worlds: the claim that they are literally present in different worlds. It will turn out that this conception has implications incompatible with the contingency view.

4. Universals existing in different possible worlds

Here is the way Lewis (Lewis, 1986a) conceives of the possibility that a universal takes part in different laws in different possible worlds. According to Lewis, universals are subject to overlap between possible worlds, in the sense that one and the same universal is part of several worlds. Lewis refutes the idea that different worlds may overlap with respect to *individuals* by arguing that this raises the paradox of accidental intrinsics (Lewis, 1986a, p. 201). An accidental property is a property an individual c has at some worlds but lacks at other worlds. An intrinsic property[15] is determined exclusively by what is the case at the spatio-temporal location of c, not by its relations to things located elsewhere. To take Lewis' example, if Humphrey himself (not Humphrey and his counterpart, but one and the same individual Humphrey) exists in two possible worlds, then the following inconsistency threatens. Having five fingers on his right hand is an accidental intrinsic property Humphrey

[15] See (Lewis, 1999, chap. 5 and 6) on the difficulty of defining that concept.

has in the actual world. It is intrinsic because it doesn't depend on anything distinct from Humphrey. And it is clearly accidental or contingent. He has it at this world W_1. But he might have had six fingers instead of five. He lacks the property of having five fingers at world W_2 where he has six fingers on his right hand. However, it is inconsistent that the same Humphrey has both five and six fingers on his right hand[16]. Therefore, Lewis concludes that the hypothesis that Humphrey exists in more than one possible world is wrong. There is no overlap between worlds with respect to individuals.

However, Lewis thinks that no such inconsistency threatens in the case of universals and that therefore there is no parallel reason pleading against the possibility that different worlds overlap with respect to universals. "A universal can safely be part of many worlds because it hasn't any accidental intrinsics." (Lewis, 1986a, p. 205, note) According to Lewis, the absence of accidental intrinsic properties makes it possible to allow for an overlap between worlds in the case of universals: "I do not see any parallel objection if worlds are said to overlap by sharing a universal. What contingent, nonrelational property of the universal could we put in place of [the] shape of the coin in raising the problem? I cannot think of any." (Lewis, 1983, p. 345, note 5)[17] Lewis says, somewhat hesitantly, that first, "there isn't much to the intrinsic nature of a universal" (*ibid.*), and second, to the extent that a universal has intrinsic properties at all, they seem to be essential to it. He thinks of such properties as being simple or composed. The extrinsic properties are contingent and change from world to world, such as the property of being instantiated N_1 times in W_1 and N_2 times in W_2. To sum up, according to Lewis, the following is true of the properties of universals: If they are intrinsic then they are essential (example: being simple or being composed) and if they are extrinsic, then they are accidental (example: the number of instantiations a universal has in a given world). If we accept the claim that universals do not have any accidental intrinsic properties, we can coherently suppose that a universal exists in more than one possible world.

What properties do universals have? Those mentioned by Lewis – being simple or composed, or being instantiated a certain number of times – cannot be its only properties. The reason is that if these were

---

[16] As Lewis notes, there is no such problem with relational properties. Humphrey can possess three dogs in W_1 and four dogs in W_2; he can be in a possession-relation to three W_1-dogs, and at the same time, without contradiction, in a possession-relation to four W_2-dogs.

[17] If a coin was present in two different possible worlds, it could be the case that it was both wholly round (in one world) and wholly octogonal (in the other world). This is the problem of accidental intrinsics for objects.

their only properties, all simple universals would be identical[18] or at least, if the number of instantiations were also taken into account, it would be impossible that two simple universals be instantiated the same number of times. However, this seems to be a quite realistic possibility: if there is an exceptionless law linking primitive universals A and B then they have the same number of instantiations. This is not question-begging in favour of the necessity of laws, for it just supposes that there is an exceptionless law that As are Bs in *one* world, e.g. in the actual one. That this is possible is not controversial.

So universals must have other properties distinguishing them. Setting aside the non-qualitative essences or quiddities discussed above, the only plausible candidates seem to be their lawful dependencies on one another. Our question is: are such relational properties accidental or essential to the universal? Someone who holds, like Armstrong and Lewis and in order to preserve modal intuitions, that they are accidental, must hold that they are extrinsic, on pains of falling victim to the contradiction of accidental intrinsics, this times concerning properties of universals.

It turns out to be sufficient to show that universals have intrinsic properties, in order to show that, if they exist in more than one possible world then those properties are essential to them (for if they exist in several possible worlds, then they cannot have accidental intrinsic properties). If there are such properties, the link between the universal and these properties has the strength of nomological necessity: it exists in all worlds where the universal itself exists. Our question becomes: Are there intrinsic properties which give rise to lawful dependencies of a universal on other universals? These properties would be essential, and so would be the links between universals, which are necessary laws. Such intrinsic and essential properties of a universal would give rise to nomological necessity in the sense given above: Truth in all worlds in which the universal exists.

5. Necessary relations between determinate and determinable universals

One intrinsic property of some universals is the property of determinate universals to be subordinate (in Fregean terminology) to their determinable universals. I shall argue that this relation is *internal*[19]

[18] I ignore the possibility of different but indistinguishable universals against which I have argued above (section 3).

[19] I shall say of relations that they are internal if and only if they supervene on their terms, and of a property that it is intrinsic if and only if it supervenes on its possessor.

and that therefore the relation of subordination between a determinate universal and its determinables is part of the essence of the determinate universal. Being an equilateral triangle is a complex universal built from the constituents: being a closed plane figure, of three sides, all sides of equal length. Being a triangle is a determinable relative to being an equilateral triangle, which is one of its determinates. E, *being an equilateral triangle*, has the second-order property of being related – in fact subordinate – to T, *being a triangle*[20]. The crucial point is that the relation of subordination between determinate and determinable universal is *internal*. An internal relation strongly supervenes on its terms; necessarily, if both of the terms exist, they are so related. In other words, in every world in which E and T exist, they are internally related so that E entails T. In those worlds, necessarily, if something is E it also is T. Moreover, the mere existence, in a world, of E, entails the existence in that world, of T. There could not be a world in which E exists but not T. Every world that contains equilateral triangles necessarily contains triangles. Taken together, these two necessary implications entail that being internally related to T is an internal relational property of E. It is not only the case that if E and T both exist, they are necessary internally related by subordination of E under T, but also that if E exists then it is necessarily internally related to T, for E's existence alone is sufficient for the existence of T. Being subordinate to T (by an internal relation) is an *essential intrinsic* property of E: It is intrinsic because the fact that E is subordinate to T does not depend on anything else than E. To be subordinate to T is also an essential property of E: E has it in every world in which it exists, because T is a constitutive part of E. (T's identity is determined by a proper part of the terms of the conjunction determining the identity of E.) Analogous arguments show that E is essentially subordinate to all determinables corresponding to any one of its constituents or to conjunctions of some (but not all) of its constituents.

Our reasoning only depends on the premise that a determinate is complex, conjunctively composed of its constituents. Nothing depends on the particular example chosen. Therefore, it can be taken to establish the following general claim on the relations between determinates and determinables:

[20]One can, following Armstrong (Armstrong, 1997, chap. 4.13), conceive of determinate universals as of complex universals resulting from a conjunctive combination of several constituents. Suppression of one of its constituents yields a universal that is determinable relatively to it. Worley (Worley, 1997) has elaborated Armstrong's account of determinate and determinable universals in a similar way.

1 It is essential (nomologically[21] necessary) for a determinate to be subordinate to each of its determinables.

I would now like to propose an argument for the necessity of one law of association, which uses 1 and the following premise.

2 It is essential (nomologically necessary) for a determinable D that each of its instantiations is also an instantiation of at least one of its determinates $D_1, \ldots, D_n$.

$$(\forall x)(\forall D)[\mathrm{Det}(D) \rightarrow \Box(Dx \rightarrow \exists D_i \in \{D_1, \ldots, D_n\} D_i x)]$$

where "Det" is the predicate "is a determinable", and $\{D_1, \ldots, D_n\}$ is the set of determinates of the determinable D.

The reason for 2 is that determinable universals exist only insofar as they are constituents of determinate universals. To be instantiated alone, the determinable would have to exist independently of all its determinates, in which case it would not be a determinable after all.

On the basis of these very general premises bearing on the relations between determinables and their determinates, I shall now argue for one case of a law of association[22] that if it is true at all, then it is necessary[23]. We may obtain this result if we add the following premise to 1 and 2.

Consider, as an example of a law of association, the Boyle-Mariotte law of ideal gases. It says that an ideal gas which has pressure P, has temperature $T = PV/nR$ (where "V" is the volume occupied by the gas, "n" is the number of moles and "R" the universal gas constant).

3 In ideal gases[24], T and P are two different determinables with respect to the same set D of determinates[25]: the set of all states of motion of the molecules composing the gas that share the mean kinetic energy specific for T and P, given a fixed volume V.

---

[21] The qualifier "nomological" means that the relation holds in all worlds in which the property exists.

[22] This argument bears only on laws of association, not on causal laws. But I shall argue later that one may generalize from this case because it would be implausible that these kinds of laws differ in their modal status.

[23] This claim must of course be distinguished from the obviously false claim that the law is a priori. Kripke (Kripke, 1972) has made a convincing case for the existence of necessary yet a posteriori truths. My thesis is that true law statements belong to this category.

[24] This restriction must be specified because T is multiply reducible; the temperature of empty space, e.g., reduces to a different property.

[25] Similarly, Hooker (Hooker, 1981, p. 497) construes the relation between a liquid's property of boiling and the underlying microscopic property of the liquid as lying on the extreme ends of a "determinate/determinable hierarchy".

From 3 alone it follows that, with the volume V fixed, the state of motion determines both P and T. Given 3 we already know that it is naturally necessary that any gas in one of the states in this set has both T and P.

The crucial question is: Is the relation between the macroscopic properties T and P that is the content of the Boyle-Mariotte law also necessary? We can derive a positive answer from our premises in the following way. By premise 2, the instantiation by the gas of the determinable property P is necessarily also an instantiation of at least one determinate property, which is in set D. By 1, each instantiation of a determinate is also an instantiation of each of its determinables, but by 3, T is such a determinable property for all states in D. Hence, an instantiation of P is necessarily an instantiation of T. On the construal of the Boyle-Mariotte law given in 3, as a relation between two determinables that have the same set of determinates, it follows from general properties of the determinate/determinable relation that the law is necessary if true. The necessity of this law is however not of a logical nature because the necessity of premise 3 is not logical. The analysis of the determination of the macroscopic properties of an object by the properties and relations of its parts is a controversial topic, but this determination is certainly not logical[26].

Without trying to argue for this claim here, it seems that an analysis along these lines is available for many laws of association between different higher-level properties of macroscopic complex objects, such as the Wiedemann-Franz law (stating the proportionality between electrical and thermal conductivity in metals) or the Dulong-Petit law (stating that the specific heat of solids has a constant universal value, which is independent of the type of solid and of the temperature within a given range). It is the empirical discovery of micro-reductions that must justify the truth of a premise analogue to 3 in each case. Therefore, the argument cannot be generalised to establish the necessity of laws of association between *fundamental* properties of microscopic particles. But if our argument is correct and if premise 3 is correct for the Boyle-Mariotte law, then we have shown that there exist necessary laws of association. And then it can be argued that it is implausible for different laws of association to differ in modal status.

What about causal laws? Causal laws are laws linking what happens at different spatio-temporal locations. Conservation laws are an important class of such laws. There are two reasons to consider that they are

[26] Armstrong (Armstrong, 1989) denies the existence of non-logical necessity. His arguments will be discussed shortly.

necessary in the same non-logical sense as laws of association. First, it is implausible to attribute a different modal status to causal laws and to laws of association for their modal status should be a consequence of their lawful status. Second, we can apply the Causal Criterion of Identity to conserved quantities. Take the conservation of mass-energy. The law of its conservation is necessary in the nomological sense that it holds wherever this quantity exists: it is constitutive for being the total energy-mass of a closed and isolated system to be conserved[27]. If some energy-like quantity of such a system is not conserved, we conclude that it is just one *form* of energy, such as potential energy or kinetic energy, but not total energy. The law of the conservation of total energy – which is a causal law in the sense that it determines what happens over time (and space[28]) – is necessary because mass-energy and other fundamental conserved quantities are conceptually linked to conservation. A property which exists in some possible world but which is not conserved is not one of them. Once again this is less than one might have hoped for. I have tried to show, not that all causal laws are necessary but only that there exist necessary causal laws[29].

Let me prevent a misinterpretation that might easily arise. We have not shown that laws are necessary in the strong sense of holding in all possible worlds. Nothing I have said prevents the existence of strange possible worlds in which there are no conservation laws. It is just that such worlds do not contain conserved quantities. The conservation of total mass-energy is necessary only in this sense: in all those worlds in which mass-energy exists, it is conserved.

---

[27]Bigelow, Ellis and Lierse argue that our actual world is the unique individual of a natural kind, and that conservation laws are grounded in this natural kind. "Conservation laws are best understood as ascribing properties to the world as a whole, properties which are essential to the natural kind to which our world belongs" (Bigelow et al., 1992, p. 385). Within this framework, the view defended here could be formulated in the following way. To be conserved is an essential property of a property of the whole world, the property of having a given total mass-energy.

[28]An energetically closed system may travel through space.

[29]If all laws were reducible to laws of association, it would not be necessary to argue separately for the existence of necessary causal laws. Against Russell (Russell, 1986) who argues that functional laws of association are the only laws, Cartwright (Cartwright, 1979) argues that causal laws cannot be reduced to such laws of association. Without trying to settle this issue here, the fact that conservation laws put constraints on the evolution of systems *over time*, whereas laws of association only constrain the properties of a system *at one time*, pleads prima facie against the possibility of a reduction of the former to the latter.

6. Incompatibilities between different determinates of one determinable

There is another source of nomological necessity: it follows from the incompatibility of different determinates of the same determinable. It is essential (nomologically necessary) for determinables that their instantiations are instantiations of only one of their determinates. In other words, different determinates exclude each other. No closed plane figure can be both triangular and quadrilateral, and no object (Lewis' example) can have both a positive and negative electric charge at the same time. In this section, I try to show that these are cases of nomological necessity, against Armstrong's (Armstrong, 1989) attempts to show that all cases of apparent nomological incompatibility can be reduced to logical or otherwise analytical necessity, and are therefore not cases of natural necessity.

According to Armstrong's combinatorial theory of possibility, any two states of affairs, *a*'s being *F* and *a*'s being *G*, are compossible if the universals *F* and *G* are entirely distinct. Armstrong's strategy to deal with apparent examples of natural necessity is to recognise the necessity of logical and other analytic relations, to count mereological relations as analytic, and then to show that wherever there are necessary relations between states of affairs, their necessity can be traced back to a logical or mereological source. He concludes that there is no genuinely *natural* but only logical necessity.

Let us see whether Armstrong can establish this reduction of nomological to logical/mereological necessity. Among the many cases of states of affairs which consist in the attribution of different determinate properties of a given determinable to one particular at one time[30], Armstrong analyses the property of mass[31]. A given particular can have only one determinate mass at a given time. Armstrong reduces the incompatibility of two states of affairs attributing two different masses to the same particular at the same time, to a mereological and thus purely analytic, not natural incompatibility. Masses are structural universals. Mereo-

[30]We can see Armstrong's analysis as a reply to Lewis' refutation of the "linguistic ersatzer", by way of showing that not everything that can be stated is a genuine possibility. "It is consistent, says Lewis, in the narrowly logical sense, to say that something is both positive and negative. [...] This seems wrong: here we seem to have an inconsistency which is not narrowly logical, but arises because positive and negative charge are two determinates of one determinable." (Lewis, 1986a, p. 154) According to Lewis, the "ersatzer" must introduce an axiom into his world-making language to prevent that the theory predicts that it is possible that a thing is both positive and negative. If such axioms are indeed necessary, it shows that modality cannot be reduced to linguistic combinations (which is what the "ersatzer" claims).

[31]Cf. Armstrong, 1989, p. 78f. Elsewhere (Armstrong, 1997, chap. 4), he applies the same strategy to duration.

logical considerations of the relation of the constituents of a structural universal to the whole explain the incompatibility of two states of affairs according to which the same particular c has both a quantity of one and of five kilograms. The explanation is that c's having a mass of five kg is equivalent to a conjunction of five states of affairs according to which five parts of c have one kg of mass each. But if the whole particular c instantiates the structural universal of having a mass of five kg, then it is necessary in the sense of "analytic" (Armstrong, 1989, p. 80) that it cannot also have the property of having a mass of one kg because this is, analytically – as a consequence of the meaning of the predicate "having a mass of five kg" – the property of one of c's proper parts. In Armstrong's words, "to attempt to combine the two properties in one thing would involve the thing's being identical with its proper part" (Armstrong, 1989, p. 79; similarly Armstrong, 1997, chap. 4.13). In Lewis' terms, no special non-logical axiom has to be introduced in order to guarantee that something cannot have both one kg and five kg of mass. The axioms of mereology suffice, if one makes the hypothesis that determinate universals are structural.

Armstrong's strategy to deal with the incompatibility of different determinates of one determinable consists in reducing the incompatibility between universals to a mereological kind of impossibility: that a whole cannot share a universal with one of its proper parts. Several objections have been raised against this analysis some of which are relevant to our topic. Let me mention two cases in which Armstrong's analysis seems to fail: the masses of fundamental particles and colours.

Armstrong's attempt to explain incompatibilities by partial identities presupposes that all those quantities that are not freely combinable are structural properties. But it is implausible that all quantities are structural: The masses of fundamental particles are not. As Menzies (Menzies, 1992, p. 733) notes, it contradicts current scientific doctrine to suppose that the masses of fundamental particles are structural properties. The properties of fundamental particles directly contradict Armstrong's claim that "if an individual has an extensive quantity, then it has parts which lie outside each other, that is which are numerically different from each other, and which go together to make up the individual and to give the individual the particular quantity that it has." (Armstrong, 1989, p. 80) The electron doesn't have its mass m_e by virtue of having two proper parts having each $1/2m_e$[32]. So why isn't it possible

[32]One might try to replace the claim that determinate universals are "structural" by the claim that they are "complex". The sense Armstrong (Armstrong, 1978, chap. 18) gives the term "structural" as applied to universals, implies that a structural universal can only

that an electron has both m_e and $1/2m_e$? Armstrong anticipates the objection that his mereological analysis may be inapplicable to incompatibilities between extensive quantities of fundamental particles. He notes that "there are grounds for thinking that, at a fundamental level, our example of mass is irreducibly intensive. For the truly fundamental particles are thought of as *point*-masses." (Armstrong, 1989, p. 80) Let us then turn to Armstrong's attempt to explain incompatibilities between intensive qualities.

Some intensive qualities, such as density, are according to Armstrong reducible to extensive quantities. Density is reducible to volume and mass, which are both extensive. "As a result, incompatibilities of density can be resolved into incompatibilities of volume and mass." (Armstrong, 1989, p. 80) But presumably (as already hinted at with respect to mass) there are also what seem to be irreducibly intensive qualities, which cannot be thus reduced to a proportion of extensive quantities. For these Armstrong proposes the strategy to apply a sort of mereological analysis based on the postulation of non-spatial parts. "Why should we not say that if science sees fit to postulate apparently irreducible intensive quantities, then what is really being postulated is the simultaneous presence of many individuals at the same place?" (Armstrong, 1989, p. 81; similarly Armstrong, 1997, chap. 4.22) If we follow Armstrong in considering the mass of fundamental particles as intensive, we could try to consider that a neutron has a proton and an electron as non-spatial parts.

I have three objections against this idea. First, it would explain just two incompatibilities (something cannot be both a neutron and a proton nor a neutron and an electron) and leave many others unexplained. It cannot, e.g., explain why nothing can be both a proton and a photon.

---

be exemplified by complex particulars (the parts of which exemplify constituents of the structural universal), which is not necessary for all determinates. (In Armstrong, 1997, chap. 3.71, he says that this is true only of "paradigm structural properties". One could also call conjunctive properties "structural" although in its case, "the constituents, the conjuncts, are properties of the very same particular that has the conjunctive property".) One could make the hypothesis that determinate universals are not structural but rather "complex" universals, which could be exemplified even by fundamental particles that have no parts. But this move would destroy the terms of Armstrong's above-mentioned argument, to the effect that the incompatibility of the exemplification of different determinates of the same determinable by one particular at one time has a purely mereological (and thus analytic) origin. It could be saved only by making the hypothesis that the particulars exemplifying complex universals have non-spatial parts even if they have no spatial parts. I discuss this hypothesis shortly.

Second, the hypothesis seems to be ad hoc[33]: Armstrong's justification to rely on mereology for explaining apparent incompatibilities was that such incompatibilities are clearly understood, in the end because they are analytical. But this certainly isn't true for a hypothetical theory that would be in some sense analogous to mereology but where non-spatial parts are combined into non-spatial wholes. No such theory has been worked out, and it seems gratuitous to rely on the hope that there could be such a theory that would provide the correct results. Third, such a theory would not be analytic in the same sense as the theorems of mereology. Even if we grant that there is a sense in which a neutron results from the "addition" of a proton and an electron, this sort of addition is not a logical operation: the properties of the resulting whole are not predictable on the grounds of logic alone, but require the knowledge of empirical laws. The masses of the "parts", the electron and the proton, do not, e.g., add up to the mass of the "whole", the neutron, according to the arithmetic law of addition but according to a more complex empirical law.

Armstrong faces a similar dilemma in the case of the incompatibility of determinate colours. In the case of colours – which Armstrong proposes to consider as extensive structural properties in the same way as mass – it is not only possible but on the contrary normally the case that the proper parts of a red object are themselves red[34]. So why cannot both the whole and one of its parts share the universal of having a mass of five kg?

Armstrong argues that colours are only phenomenologically simple but can be reduced to structured physical properties. Taking up a suggestion of the Tractatus (6.3751), Armstrong (Armstrong, 1989, pp. 82–84) holds that secondary qualities such as colours are to be identified with "primary-quality structures" (Armstrong, 1989, p. 83). He suggests that incompatibilities between the latter can always be reduced to incompatibilities between extensive quantities. But he doesn't show this in detail for any secondary quality. What he does instead is show how this strategy works for explaining the incompatibility between different velocities, and then declare without further argument that the same strategy works for secondary qualities. But it seems that if colours, e.g., can be reduced to complex physical structural properties, such as the capacity of reflecting light of certain wavelengths, these reducing prop-

[33]Only in this context does Armstrong consider such a possibility. Elsewhere, he takes it as obvious that "two material objects cannot be at the same place at the same time." (Armstrong, 1968, p. 240)
[34]Cf. (Macdonald, 1991, p. 162).

erties are intensive. The impossibility of a photon's being both of the wavelength 500 Å and 1000 Å, does not stem from its being composed of two parts with a wavelength of 500 Å each, short of making the doubtful hypothesis of non-spatial parts (Cf. Menzies, 1992, 733)[35],[36].

In the end, after this look at Armstrong's suggestion to account for incompatibilities between intensive qualities in terms of a speculative non-spatial mereology, we arrive at the conclusion that these incompatibilities are irreducible to logical or mereological incompatibilities. Against both Armstrong and Lewis, I suggest that these incompatibilities are instances of natural necessity, which have the same origin as necessary laws of nature, for which we have argued above[37].

---

[35]My objection is that Armstrong's strategy to show that the incompatibility is grounded on a partial identity between the incompatible states of affairs, and thus on a mereological incompatibility, doesn't work. Bradley (Bradley, 1989, p. 36) objects that it is ad hoc: I think that this objection is justified only where Armstrong applies it to properties for which our best current scientific theories gives us grounds for thinking that they are simple. It is ad hoc to overrule science and simply postulate that there must be hidden structure in order for there to be a solution to a difficulty encountered by the philosophical theory. This looks like, in Lewis words, "letting philosophy dictate to science" (Lewis, 1992, p. 212). But with respect to colours, the objection seems misdirected for here science does give us grounds for thinking that colours are complex properties.

[36]Wittgenstein (Wittgenstein, 1966, p. 35) considers a similar analysis of intensive qualities as conjunctions of their parts, and rejects it for reasons similar to those indicated in the text.

[37]Several authors have defended the thesis that laws of nature are necessary although not always in the sense intended here, of holding in all worlds that share our actual universals, and for reasons different from those presented in this paper. Cf. Shoemaker, 1980, Shoemaker, 1998, Swoyer, 1982, Fales, 1993, Bigelow et al., 1992, Ellis and Lierse, 1994, and Ellis, 1999, Ellis, 2000, Ellis, 2001. This is not the place for a detailed analysis of the differences between the accounts offered by these authors and mine (Cf. Kistler, 2002). Let me just note that the position defended here differs from Ellis' and Lierse's "dispositional essentialism" (DE) in several important respects. First, according to DE, laws are grounded in dispositions that are essential properties of *natural kinds*, which are primitive and fundamental kinds of entities. On the view defended here, it is the essential nomological properties of *properties* (here construed as universals) that provide the grounding of laws. My main reason for holding that (simple) properties are more fundamental than natural kinds is that kinds are complex types of substances, which share structural properties. But the constituents of the structure are held together in virtue of laws governing those constituent properties. So it seems that the identity of a kind depends on the identity of its constitutive properties. Second, the fundamental essences of DE are *causal powers* belonging to natural kinds. I have argued more generally for the existence of essential *nomological properties* of properties, of which causal powers are only a special kind. Third, according to DE, if a disposition $\langle C, E\rangle$ to have the effect E in circumstances C is "causally determinate", then "an event of the kind E must occur to x [...] as a result of a C-type event occurring to x at t" (Ellis, 2001, p. 130). However, perfectly deterministic dispositions do not obey this condition because their effects are typically *themselves dispositional* and do not always manifest themselves in a way that only depends on C. For example, a negative electrical charge at point P has the disposition (in virtue of a deterministic law) to create an electrical field that has, at some point Q distant from P, the strength E. But if, as will generally be the case, the charge is not the only one around, the total electrical field strength at Q will not be E, as determined by the charge at P; the total field strength will result rather from the superposition of many dispositions for an electrical field at Q.

7. Conclusion

Starting from the idea that laws are second-order relations between properties, and thus equivalent to second-order relational properties of properties, I have argued for the thesis that at least some of these nomological properties of properties are essential to them, in the sense that the first-order property would not be the property it is if it did not possess the second-order nomological property. If this is true then the laws corresponding to these nomological properties are necessary in a particular sense: although they do not hold in all possible worlds – they are not logically necessary – they hold in all worlds in which the first-order properties exist.

To establish the thesis of the nomological necessity of at least some laws, I have first argued against two important ways of justifying a crucial requirement for the opposite thesis: if the laws of nature were contingent, the universals taking part in them would have to have an essence independent of their lawful relations to other universals. I have tried to show that Rosenkrantz' "haecceities" are not tied to criteria of identity which could ground their identity and difference across different possible worlds. Furthermore, I have shown Cover and O'Leary-Hawthorne's argument for the existence of "quiddities" of universals, based upon the alleged possibility of the existence of indistinguishable universals, to be invalid.

Second, I have given two positive arguments for the necessity of at least some laws of nature. The first concerns laws of association linking different properties that are instantiated at the same time at the same place. If a law linking macroscopic properties such as the Boyle-Mariotte law linking the temperature, pressure and volume of an ideal gas can be construed as linking different determinables of the same class of determinates, then the logic of the relations between determinates and determinables allows to establish that it is necessary if it is true.

The second argument regards the impossibility of several determinates of the same determinable to be instantiated by the same particular at a given time. I have tried to show that such incompatibilites require the postulate of a specific nomological type of necessity, against Armstrong's argument that it can be reduced to analytic (more precisely mereological) necessity. Insofar as this necessity is not analytic, it gives us a reason to postulate nomological necessity as a fundamental kind of necessity.

As we have seen in the beginning of the paper, Lewis' Humean combinatorialism allows laws of association to be necessary insofar as they constrain relations between different aspects of events, in other words because they constrain only what is the case at one spatio-temporal location. Although Lewis (Lewis, 1986a, p. 114) thinks that we have neither reasons for nor against the hypothesis that laws of association are necessary, the latter thesis is compatible with Lewis' Humean combinatorialism, which postulates independence only between what happens at different space-time regions. However, Armstrong's stronger Tractarian combinatorialism requires that even relations between facts obtaining at the same spatio-temporal location are contingent except if they are reducible to logical or otherwise analytic relations. Our conclusion, that laws of association are necessary, is therefore incompatible with Armstrong's metaphysical framework. Furthermore, our conclusion that causal laws, such as laws of conservation, are necessary is incompatible with both Lewis' and Armstrong's metaphysics.

Acknowledgements. I thank my auditors in Copenhagen for their critical comments on an earlier version of this paper, and Joan Cullen for her help in improving the final draft.

References

Adams, R. M. (1979). Primitive thisness and identity. *Journal of Philosophy*, 76:5–26.

Armstrong, D. (1968). *A Materialist Theory of the Mind.* Routledge, London.

Armstrong, D. (1978). *Universals and Scientific Realism*, volume: *A Theory of Universals.* Cambridge University Press, Cambridge.

Armstrong, D. (1983). *What is a Law of Nature?* Cambridge University Press, Cambridge.

Armstrong, D. (1984). Replies. In Bogdan, R. J., editor, *D. M. Armstrong, Profiles*, volume 4, pages 225–269, Dordrecht. Reidel.

Armstrong, D. (1989). *A Combinatorial Theory of Possibility.* Cambridge University Press, Cambridge.

Armstrong, D. (1997). *A World of States of Affairs.* Cambridge University Press, Cambridge.

Bacon, J., Campbell, K., and Reinhardt, L., editors (1993). *Ontology, Causality and Mind. Essays in Honour of D. M. Armstrong*, Cambridge. Cambridge University Press.

Bigelow, J. (1999). Scientific ellisianism. In Sankey, H., editor, *Causation and Laws of Nature*, pages 56–76, Dordrecht. Kluwer.

Bigelow, J., Ellis, B., and Lierse, C. (1992). The world as one of a kind: Natural necessity and laws of nature. *The British Journal for the Philosophy of Science*, 43:371–388.

Black, M. (1952). The identity of indiscernibles. *Mind*, 61:153–164.

Bradley, R. (1989). Possibility and combinatorialism: Wittgenstein versus Armstrong. *Canadian Journal of Philosophy*, 19:15–41.

Cartwright, N. (1979). Causal laws and effective strategies. *Noûs*, 13:419–437. (Reprinted in Cartwright, 1983).

Cartwright, N. (1983). *How the Law of Physics Lie.* Clarendon, Oxford.

Dretske, F. (1977). Laws of nature. *Philosophy of Science*, 44:248–268.

Ellis, B. (1999). Causal powers and laws of nature. In Sankey, H., editor, *Causation and Laws of Nature*, pages 19–34, Dordrecht. Kluwer.

Ellis, B. (2000). The new essentialism and the scientific image of mankind. *Epistemologia*, 23:189–210.

Ellis, B. (2001). *Scientific Essentialism.* Cambridge University Press, Cambridge.

Ellis, B. and Lierse, C. (1994). Dispositional essentialism. *Australasian Journal of Philosophy*, 72:27–45.

Fales, E. (1993). Are causal laws contingent? In Bacon et al., 1993.

French, S. and Redhead, M. (1988). Quantum physics and the identity of indiscernibles. *The British Journal for the Philosophy of Science*, 39:233–246.

Hooker, C. A. (1981). Towards a general theory of reduction. *Dialogue*, 20:38–59, 201–236, 496–529. Part I: *Historical and Scientific Setting*, Part II: *Identity in Reduction*, and Part III: *Cross-categorical Reduction.*

Kistler, M. (2002). The causal criterion of reality and the necessity of laws of nature. *Metaphysica*, 3:57–86.

Kripke, S. (1972). Naming and necessity. In Harman and Davidson, editors, *Semantics of Natural Language*, Dordrecht. Reidel.

Lewis, D. (1973). *Counterfactuals.* Basil Blackwell, Oxford.

Lewis, D. (1983). New work for a theory of universals. *Australasian Journal of Philosophy*, 70:211–224.

Lewis, D. (1986a). *On the Plurality of Worlds.* Basil Blackwell, Oxford.

Lewis, D. (1986b). *Philosophical Papers*, volume II. Oxford University Press, Oxford.

Lewis, D. (1992). Review of Armstrong, 1989. *Australasian Journal of Philosophy*, 70:211–224.

Lewis, D. (1994). Humean supervenience debugged. *Mind*, 103:473–490.

Lewis, D. (1999). *Papers in Metaphysics and Epistemology.* Oxford University Press, Oxford.

Lewis, D. and Langton, R. (1998). Defining 'intrinsic'. In Lewis, 1999.

Lowe, E. J. (1998). Entity, identity and unity. *Erkenntnis*, 48:191–208.

Macdonald, C. (1991). Review of Armstrong, 1989. *Philosophical Books*, 32:163–164.

Mellor, D. H. (1993). Properties and predicates. In Bacon et al., 1993, pp. 101–113. (Also in: Mellor, D.: *Matters of Metaphysics*. Cambridge University Press, Cambridge).

Mellor, D. H. (1997). Introduction. In Mellor, D. H. and Oliver, A., editors, *Properties*, pages 1–33, Oxford. Oxford University Press.

Menzies, P. (1992). Review of Armstrong, 1989. *Philosophy and Phenomenological Research*, 52:731–734.

O'Leary-Hawthorne, J. and Cover, J. A. (1997). Framing the thisness issue. *Australasian Journal of Philosophy*, 75:102–108.

Redhead, M. and Teller, P. (1992). Particle labels and the theory of indistinguishable particles in quantum mechanics. *The British Journal for the Philosophy of Science*, 43:201–218.

Rosenkrantz, G. S. (1993). *Haecceity : An Ontological Essay*. Kluwer, Dordrecht.

Russell, B. (1986). On the notion of cause. In *Mysticism and Logic*, London. Routledge.

Shoemaker, S. (1980). Causality and properties. In *Identity, Cause, and Mind*, Ithaca (N.Y.). Cornell University Press.

Shoemaker, S. (1998). Causal and metaphysical necessity. *Pacific Philosophical Quarterly*, 79:59–77.

Swinburne, R. (1995). Thisness. *Australasian Journal of Philosophy*, 73:389–400.

Swoyer, C. (1982). The nature of natural laws. *Australasian Journal of Philosophy*, 60:203–223.

Tooley, M. (1987). *Causation: A Realist Approach*. Clarendon Press, Oxford.

Wittgenstein, L. (1966). Some remarks on logical form. In Copi, I. M. and Beard, R. W., editors, *Essays on Wittgenstein's Tractatus*, London. Routledge and Kegan Paul.

Worley, S. (1997). Determination and mental causation. *Erkenntnis*, 46:281–304.

LAWS OF NATURE – A SKEPTICAL VIEW

Uwe Meixner

Let me begin by making

1. Some general skeptical remarks regarding laws of nature

The epistemologically unproblematic position on laws of nature is the following. There are regularities in nature, more or less adequately describable by general sentences. Some of these regularities we are particularly interested in, because of the systematizing function the statements describing them can exercise in the formulation of our theories of nature, and because these statements, due to their simplicity, are found to be explanatory by us and can be used in the explanation of a wide range of natural phenomena. Regularities in nature with a high systematizing and explanatory power for us – or rather, the statements describing them, which properly speaking have this power for us – we traditionally call "laws of nature." That's all there is to laws of nature.

The epistemologically problematic position on laws of nature, however, is the following. Behind some regularities in nature there are form-like ontic principles (form-like *archai*) that determine these regularities and confer necessity on them. Moreover, except for the workings of absolute chance, those form-like ontic principles determine just about everything in the world, including its very existence. They are the laws of nature.

It is surprising that this piece of ancient Platonism is found to be attractive by so many modern thinkers. What could make one believe in it? It is clear from the start that there can be no proof for the existence of laws of nature in the epistemologically problematic sense. Hence, if proof is required for the rationality of belief, the belief in laws of nature in the epistemologically problematic sense is quite irrational. But perhaps something less than proof is required for the rationality of belief (although, somewhat unfairly, the usual exception is made with respect to belief in God). Perhaps a plausible argument – for example, an *argument to the best explanation* – is all that is needed. And indeed,

J. Faye, P. Needham, U. Scheffler and M. Urchs, (eds.), Nature's Principles, 229–238.

laws of nature in the (it seems now, merely *prima facie*) epistemologically problematic sense are said, by not a few philosophers, to be the best explanation of certain regularities in nature. Can it be, they ask (rhetorically), that those regularities are here by mere chance? Who can believe this? And so and so on, till the desired conclusion is reached.

But this argument for the existence of laws of nature in the epistemologically problematic sense, which is strangely reminiscent of the teleological argument for the existence of God, can be of interest only to those philosophers that believe that some regularities in nature stand in need of explanation - an explanation that consists in more than in deducing them from more basic regularities. And there seems to be nothing irrational in not requiring such an explanation – especially in view of the fact that we need to stop asking for explanation at some point anyway. But if we *do* ask for explanation in the case at hand, by what could we be made to think that laws of nature in the epistemologically problematic sense are the best explanation of certain regularities in nature? I fear, by nothing except a very large piece of begging the question: by considering every explanation of the regularities that does not invoke laws of nature in the problematic sense to be automatically less good than the explanation that does invoke them. In philosophy, alleged arguments to the best explanation usually turn out to be arguments to the metaphysically *best-liked* (the metaphysically *most beloved*) explanation, and here we apparently have a fine example of this. But I will not rest with these very general remarks, but shall take a closer look at the epistemology of laws of nature.

2. Bas van Fraassen and TAD (Tooley-Armstrong-Dretske)

Before presenting my own skeptical argument regarding laws of nature, I will examine the skeptical argument Bas van Fraassen has directed specifically against Michael Tooley's, David Armstrong's and Fred Dretske's, in short: TAD's, views on the nature of laws of nature. It turns out that we can abstract from the specificities of the intended target and can take van Fraassen's argument as being quite generally directed against an objective conception of laws of nature that in some way or other involves the idea of *necessity*.

Van Fraassen basically presents his case against TAD on pp. 94–99 of *Laws and Symmetry*. What he says there is less than clear. But he seems to have the following in mind:

1 TAD thinks that the form of a sentence expressing a simple law of nature is "F necessitates G," where F and G are first-order

universals and necessitation is a logically contingent and objective (second-order) relation between them.

2 TAD thinks that "*F* necessitates *G*" logically implies (or entails) "All *F* are *G*," but not vice versa.

3 Question: Which relation that satisfies all the constraints contained in 1 and 2 is the relation of necessitation that TAD has in mind?

4 There is no satisfactory answer to 3.

This is a skeptical argument. The conclusion it argues for is that we just don't know – and that TAD doesn't know either – *what* TAD means by "necessitation" and "to necessitate" (or whatever expression is used). If this is correct, then TAD's account of laws of nature turns out to be quite unsuccessful.

Prima facie it seems very easy to refute the skeptical argument: Simply define "*F* necessitates *G*" to mean the same as "It is (objectively) necessary that all x that are (have) *F* also are (have) *G*," or, alternatively, as meaning the same as "For every x it is (objectively) necessary that if it is (has) *F*, it also is (has) *G*." The first definition, at least, seems to provide a very clear answer to the question formulated in 3.[1]

But it is not a definition that TAD would or should allow. For one thing, it takes all the (comparative) novelty away TAD has modestly claimed for his account of laws of nature: it becomes an ordinary modal, *necessitarian* account (as van Fraassen calls such an approach; see van Fraassen, 1989, p. 65). And, true, on a standard logic of "necessary" the constraint in 2 is satisfied if this definition is used. But one can well ask: what, *specifically*, is this necessity that turns up in the definiens? It seems no proper explication of "necessitation" and "to necessitate" has been effected at all, merely a synonym offered, whose only advantage over the original expression is that by giving us a bit more logical structure than the original has, it makes clear *how* necessitation could fulfill the constraint in 2. The real work is still undone: to specify the *right* kind of necessity, which must be a *logically contingent* and *objective* necessity.[2]

---

[1]The second definiton is a little less clear, at least to some minds, since it involves *de re* necessity.

[2]While indicating this kind of objection, Armstrong also offers a quite different reason against the modal, necessitarian analysis of *necessitation*, a reason that involves the "Paradoxes of Confirmation." See Armstrong, 1983, pp. 87–88. I do not think that Armstrong's reason is a serious reason, because the problem of the paradoxical confirmation of lawhood Armstrong points out may well be taken not to be a problem for the modal analysis of necessitation, but a problem for confirmation theory.

The required logical contingency of necessitation (i. e., that in at least some of its instantiations it is not instantiated by logical necessity) could be taken care of by simply defining "F necessitates G" to mean the same as "all F are G." This definition would also circumvent all difficulties that may be connected with the concept of logically contingent necessity. But it would leave us with an interpretation of necessitation that is *surely not* the interpretation that TAD, or anyone else, has in mind.[3]

Well, what is this interpretation? What is the interpretation of necessitation TAD has in mind? Here is what one of TAD's spokesmen has to say on this account:

> [T]he inexplicability of necessitation just has to be accepted. Necessitation, the way that one Form (universal) brings another along with it as Plato puts it in the *Phaedo* (104d - 105), is a primitive, or near primitive, which we are forced to postulate. (Armstrong, 1983, p. 92.)

Quite obviously, we cannot explicate all concepts at once, and perhaps some concepts are inexplicable in any system of concepts available to us. But if we arrive at a concept that we do not or cannot explicate, then we should at least be able to give some indications of its contents. If not, the concept is a complete *nothing* for us, and all we really have before us is an empty word without legitimate use. Unfortunately, this is just what "necessitation" and "to necessitate" seem to be (at least in TAD's mouth): empty words. The historical reference Armstrong provides is not helpful at all, since in the cited passage Plato is quite clearly speaking about a broadly logical relation: the bringing-along that holds between *being colored* and *being extended*, for example. That relation is well understood, but it is not the relation Armstrong or TAD presume to mean when they talk of necessitation – that is, of some *logically contingent* relation – as being constitutive for simple laws of nature.

Perhaps, in view of this, it is best to return to the modal, necessitarian account – notwithstanding TAD's protests – and make a serious effort to elucidate the concept of necessity that is invoked when we say that "F necessitates G" (assuming this to be the form of a sentence expressing a simple law of nature) just means that *necessarily* (in an objective sense) all F are G.

In van Fraassen, 1989, chapter 4, van Fraassen also canvasses the necessitarian approach to laws of nature (the approach that TAD's approach simply reduces to, once the above analysis of "necessitates" is accepted, and accepting it, as I said, seems the best thing to do after

---

[3]It is excluded by the "not vice versa" in 2.

all). Van Fraassen finds the necessitarian approach wanting mainly on account of the realism about possible worlds that seems to be implied by it. But the problem with the necessitarian approach appears to me to be of a much more elementary nature. It is essentially the same difficulty as the one that was pointed out in the above argument against TAD. That argument had the conclusion that we just don't know what "necessitation" means. Now, that conclusion is not at all removed, it is just moved one step backward, if we leave TAD where he stands and point out that necessitation is *necessary extensional inclusion*, where the inclusion is "passive," i. e., "F necessitates G" is taken to be definitionally equivalent to "It is necessary that F *is extensionally included* in G," and where the "necessary" is taken to refer to some logically contingent and objective necessity. For we just don't know *which* concept, exactly, *is* that necessity. Again, we are left with a word that seems condemned to emptiness by the very constraints put on its interpretation: "(*objectively*) necessary (*but not logically necessary*)" – a word that is no less empty if we assume a general logic for it (S4 or S5, or whatever).

Things seem to brighten up for a moment when we add the word "nomologically" to the word "necessary": "nomologically necessary." Yes, this seems to indicate precisely the kind of necessity that is needed for analyzing necessitation – the relation that is constitutive for simple laws of nature. But we are laboring under an illusion. For reflect: In order to know what *nomological* necessity is, we need to know what a law of nature is. But we haven't found that out yet; in fact, we are trying to find it out via finding out what nomological necessity is.[4] The whole move is entirely hopeless.

3. We don't know which regularities are the laws of nature

The skeptical potential in van Fraassen's argument is considerable – especially if we free it from the particularities of its intended target (i. e., the ideas on laws of nature that are peculiar to TAD) and give it wider implications. For skeptical purposes, just like Hume's argument regarding *causation*, it exploits the fearful philosophical difficulty of specifying (*truly* specifying, and not just making words about it) a necessity with normal logical properties (mainly, $\Box A \supset A$, without $A \supset \Box A$) that is at once objective and yet very different from logical necessity.

The following skeptical approach to laws of nature, however, is quite different from the one described above: necessitation, whether analyzed

[4]This, in a nutshell, is the criticism that I offer in Meixner, 1997.

by making use of a concept of necessity or regarded as a primitive, plays no role in it at all, nor does necessity, nor does any particular view about the form of sentences that express simple laws of nature. Moreover, while the above-described skepticism was an instance of *meaning skepticism*, what follows will be an instance of epistemological skepticism in a narrower sense.

I begin by positing

Thesis 1 *Any world w which has the same laws of nature as the real world, w^*, cannot be justifiedly distinguished by its inhabitants from any world w' that is phenomenally*[5] *identical with w, but is merely phenomenally compatible with the laws (of nature) of w^* and has different laws of nature than w^*.*

For suppose w is a world in which the same laws of nature as in w^* hold; and suppose that w' is a world that is phenomenally identical with w, but is merely phenomenally compatible with the laws of nature of w^* and has different laws than w^*. Clearly, w cannot be justifiedly distinguished by its inhabitants from w' (on what grounds could they do so?); for all they know, w is identical with w'.

Thesis 1 has the following obvious corollary:

Thesis 2 *The real world, w^*, cannot be justifiedly distinguished by its inhabitants from any world w that is phenomenally identical with w^*, but is merely phenomenally compatible with the laws of nature of w^* and has different laws than w^*.*

This corollary of Thesis 1, in turn, has the following consequences. Suppose that w_1, w_2 and w_3 are worlds which are phenomenally identical with the real world, w^*, and therefore phenomenally compatible with the laws of nature of w^*. But in w_1 there are laws of nature in addition to those in w^*; in w_2, on the contrary, the set of laws is a proper non-empty subset of the set of laws of w^*; in w_3, finally, there are no laws of nature at all. Now, according to Thesis 2, the real world, w^*, cannot be justifiedly distinguished by its inhabitants (and this means, in particular, by us) from w_1, w_2 and w_3. For all they know, w^* is any one of these three worlds. How, then, can they be justified in assuming that the set of the laws of nature of the real world comprises precisely

[5]The notion of the *phenomenal* is here to be taken in an ontological, not in an epistemological sense: in a sense in which, for example, microphysical facts, states and events are *phenomena*, even though they are not directly observable. In this sense, the phenomenal facts (states, events) are precisely the non-modalized facts (states, events): the facts (states, events) that do not involve any modality (alethic or non-alethic).

those items that they have hit on in pursuing their scientific enterprises? Suppose they are lucky and have indeed exactly the laws of nature of the real world – that is: *the* laws of nature – in the set of principles they have hit on. But w^* cannot be justifiedly distinguished by them from w_1; therefore, the set of the laws of nature of w_1 must be as good a candidate for them for being the set of the laws of nature of w^* as the set they have hit on. And w^* also cannot be justifiedly distinguished by them from w_2; therefore, the set of the laws of nature of w_2 must in its turn be as good a candidate for them for being the set of the laws of nature of w^* as the set they have hit on. Finally, w^* cannot be justifiedly distinguished by them from w_3; therefore, the set of the laws of nature of w_3, *the empty set*, must again be as good a candidate for them for being the set of the laws of nature of w^* as the set they have hit on. This means: *they really do not have any justified opinion as to which items are the laws of nature of the real world*, even if they are so lucky as to have, in their scientific enterprises, hit on precisely the principles which are in fact the laws of nature of the real world. For all they *know*, there might even be no laws of nature (of the real world) at all.

This skeptical argument is obviously based on the proliferation of worlds which are phenomenally identical with the real world and therefore phenomenally compatible with the laws of nature of the real world, but which nevertheless have laws of nature *differing* from those of the real world. The only way to block this proliferation in such a manner as to make the skeptical argument impossible is to postulate *that a world which is phenomenally identical with the real world, and thus phenomenally compatible with the laws of nature of the real world, has the very same laws of nature as the real world.* But this postulate will not help us if the only reason to believe in it is that it allows us to escape skepticism with respect to laws of nature.

It seems, however, plausible on *independent* grounds that worlds which are phenomenally identical to each other are *simpliciter identical.* If this is true, then the above postulate falls out as a trivial consequence and skepticism with respect to laws of nature is avoided. Yet, on closer examination, the stated identity principle for worlds becomes suspect. It implies that the laws of nature of the real world are *completely determined* by the phenomena of the real world.[6] Can this be true? Only if *every eligibile* regularity found in nature is (or stands for) a law of

[6]According to the stated identity principle for worlds, it cannot be that we have these very same phenomena but laws that are different from the actual laws. For if this could be, then there would be a world which is phenomenally identical with the real world, but different from it (since it has different laws). But, on the identity principle for worlds now under consideration, there is no such world.

nature, or, indeed, *no such regularity*. If the phenomena of the real world completely determine its laws of nature, then there is no reason why they should determine the actual and eligible regularity R to be a law of nature, but not the equally actual and eligible regularity R'. But haven't we all been taught that not all eligible regularities found in nature are laws of nature, but only *some* such regularities?

However, there appears to be yet another way to justify the above postulate. One could *stipulate* the laws of nature of a world that is phenomenally compatible with the laws of nature of the real world to be precisely those features of it that are common to all worlds that are phenomenally compatible with the laws of nature of the real world. Then, by stipulation, a world which is phenomenally compatible with the laws of nature of the real world automatically has the very same laws of nature as the real world, and there is no longer any logical gap between *being phenomenally compatible with the laws of nature of the real world* and *having the same laws of nature as the real world.*

This stipulation requires that any particular instance of a general law of nature of a world w which is phenomenally compatible with the laws of nature of the real world is also a law of nature of w: with $\forall x(Fx \supset Gx)$ being a law of nature of it, $Fa \supset Ga$ must also be a law of nature of it, since if the former feature is common to all worlds phenomenally compatible with the laws of nature of the real world, then the latter certainly is so, too. This consequence is contrary to the usual conception of laws of nature as general regularities. But more importantly, it is unclear what could be a rational motivation for the suggested stipulation – besides the motivation to close the logical gap mentioned above (which motivation, by itself, doesn't count much). Finally, the suggested stipulation still tells us nothing at all about which principles are the laws of nature of the real world. *If* we accept it, we can indeed safely conclude that a world phenomenally identical to the real world has the very same laws of nature as the real world. But this conclusion is still compatible with the laws of nature (of the real world) being so and so, or rather such and such, and, most disquietingly, it is compatible with there being no laws of nature at all. The phenomena of the real world leave all these possibilities completely open.

Thus we find ourselves caught in a dilemma: The very concept of a law of nature demands that such laws transcend the phenomena (and therefore: which principles are laws of nature, and which are not, is not completely determined by the phenomena). But this transcendence, on the other hand, as we have seen, puts laws of nature outside of our epistemic reach. In this respect, the concept of law of nature is strikingly like the concept of God. And, indeed, in atheistic metaphysics the former concept functions in many respects just like the latter: the concept of law of nature has replaced the concept of God. It is appropriate to quote Wittgenstein here:

> The whole modern conception of the world is founded on the illusion that the so-called laws of nature are the explanations of natural phenomena.
>
> Thus people today stop at the laws of nature, treating them as something inviolable, just as God and Fate were treated in past ages.
>
> And in fact both are right and both wrong[7]: though the view of the ancients is clearer in so far as they have a clear and acknowledged terminus, while the modern system tries to make it look as if everything were explained. (Wittgenstein, 1961, 6.371 and 6.372)

There is, of course, a way to escape from the dilemma that has just been described. One *can* have the transcendence of laws of nature over the phenomena *and* keep laws of nature within our epistemic reach. But only if the status of law of nature is conferred *by us*, is our making, is relative to our beliefs and decisions, and hence can also be taken away by us.

I have come back to the epistemologically unproblematic position on laws of nature I started out with, with the addition of a philosophical argument for it. Yet, it must be conceded that laws of nature are *normally* intended to be more than what the unproblematic position allows them to be. The very expression "law of nature" demonstrates this fact, and the above quote from Wittgenstein effectively underlines what the modern mind more or less consciously expects from laws of nature: a rational substitute, so it believes (consciously or not), for God. But, as I hope to have made clear, it is epistemological foolhardiness, and far from rational, to believe in laws of nature in the epistemologically problematic sense, since nobody can know which items are the laws of nature in this sense.

4. Hume's Dream

Let me close by recounting a philosophical story, totally apocryphal of course.

David Hume dreams that he comes into a gigantic hall where he has never been before, the floor of which is covered by a huge carpet. But only a small portion of the carpet can be seen, displaying a very beautiful

[7]Why are they both right and both wrong? Presumably Wittgenstein is suggesting that the urge for explanation that motivates both the ancient and the modern view is natural and somehow valuable, and in this sense "right," but that it is nevertheless (since it is ultimately a *metaphysical* urge that aims at saying what cannot be said) philosophically misguided, and in this sense "wrong."

pattern. The rest of the carpet is concealed by a white sheet, which, as Hume quickly finds out, cannot be removed. Hume, in his dream, looks at the pattern on the portion of the carpet he can see, and announces to someone he knows is waiting for an answer (he has to answer the question "What do you know about the pattern of the carpet covering the floor of this hall?" and is allowed only two attempts; he vaguely feels that something important depends on his answer, but doesn't know what it is): "This pattern here displayed is the pattern of the *entire* carpet covering the floor of this hall." There is no response. Hume, dismayed, is not sure what this is supposed to mean; perhaps the answer was wrong, perhaps not sufficient. After anxiously staring at the revealed portion of the carpet a bit longer, he announces with regained confidence ("This must be it!"): "Even better, there can be no carpet fit to cover the floor of this hall that does not in its entirety have the pattern that is here displayed." Barely are these words out, when Hume hears a voice from nowhere pronouncing calmly and distinctly, "You have answered very foolishly," which for some reason frightens him so much that he wakes up with a start. From this moment on, so the story goes, Hume really started *to think* (with the results so well known, and so often not heeded), although he quickly forgot all about his dream: the hall, the carpet, and the voice.

References

Armstrong, D. M. (1983). *What is a Law of Nature?* Cambridge University Press, Cambridge.

Dretske, F. (1977). Laws of nature. *Philosophy of Science*, 44:248–268.

Meixner, U. (1997). *Ereignis und Substanz.* Schöningh / mentis, Paderborn.

Tooley, M. (1997). The nature of law. *Canadian Journal of Philosophy*, 7:667–698.

van Fraassen, B. C. (1989). *Laws and Symmetry.* Clarendon, Oxford.

Wittgenstein, L. (1961). *Tractatus logico-philosophicus.* Routledge and Kegan Paul, London.

THE LAWS' PROPERTIES

Johannes Persson

To give a new theory a run for its money the universe one assumes might be filled up with new properties. The strategy reflects a possible and frequently presupposed ontological attitude amongst scientific realists: new science → new laws → new properties. But many of us are convinced that there is an important distinction to be made between concepts and properties and therefore between laws of different ontological status. New science → new concepts (whether or not *via* the laws) is an innocent step while new concepts → new properties is sometimes justified but at other times doubtful. Not even the hardcore realist would like to have to accept that each concept in true theories marks a real property. Conversely, those of us who are not convinced realists in general still want to accept that many scientific concepts mark properties. There is need for a good look at the laws' properties.

1. What Properties?

The connection between properties and laws, when a new property is discovered, as opposed to when only a new concept has surfaced, must first be expanded on. There are many interpretations of properties, ranging from various versions of nominalism to different kinds of realism. They are not all suitable for this task.

Most clearly, nominalistic interpretations of properties are problematic in this context. Predicate- and concept-nominalists deny the distinction between concepts and properties. They cannot even approach the issue above. But also other nominalisms, such as class- and resemblance-nominalism, are poorly adapted. The mix between realism and anti-realism we are sometimes interested in presupposes the possibility of having properties in the world at the same time as having something only superficially mirroring these properties. Both properties and fake properties are needed. According to all variants of nominalism, however, properties are never in the world but always accounted for in some other way, i.e. all properties are fake properties.

J. Faye, P. Needham, U. Scheffler and M. Urchs, (eds.), Nature's Principles, 239–254.

The other traditional position, realism, is a better choice. But most familiar realist views face other obstacles. A theory like Platonism or Aristotelianism typically takes universals to be related by the laws. They often make laws relating properties second-order universals:

1 $L(P, Q)$

Many familiar laws in the sciences seem instead to relate objects or individuals. Newton's first law of motion says:

> Every body will continue in its state of rest, or of uniform motion in a right [i.e. straight] line, unless it is compelled to change that state by forces impressed upon it (Newton's *Principia*, p. 13).

This is a law not like 1 but rather of the following form:

2 Objects with property P affect objects with property Q

Sometimes we think that expressions like 1 entail expressions like 2. To see that this is not so one needs only to think of the possibility that laws are combined in some situations. Many objects have both a mass and a charge. They are then affected by both the laws $F = MA$ and $F = C_1C_2/d^2$. As a consequence, the objects display another behaviour of type 2 than any of the laws of type 1 taken by itself entails. There is thus a general difficulty in moving between laws of level 1 and 2.[1]

I prefer instead a view of properties that seems to have neither of the problems these familiar property-views have. A trope view neither denies the distinction between concepts and properties nor does it imply, as we will see, that laws cannot be of first-order. Tropes are presently in vogue but despite their popularity theories of them are neither better nor more uniformly developed than more traditional theories of properties. So I will not build on any specific trope view here. Instead the essentials of the kind of theory I rely on will be presented below.

1.1 Tropes, universals, and bundles

Tropes are once-for-all occurring properties (Williams, 1953, p. 172). The key to understanding tropes is to contrast properties as kinds with properties as instances – and to think of the latter as the fundamental ontological category. Metaphysicians have often thought the kind to be fundamental; and often conceived of instances as primarily instances of

[1] It can of course be doubted whether 2 is a good way of expressing the laws. Perhaps they should rather always be of type 1? Max Kistler (Kistler, 1999) provides a good case for this being so, and accepts the consequence that laws do not govern the behaviour of objects.

a kind. Fortunately, one has only to turn to the epistemologists of the British empiricist tradition to find a helpful analogue from the opposite perspective. Hume, for instance, discusses impressions in a way that is congenial to a trope approach. While no impressions become blue by being related to other impressions, the general or abstract notion of blue presupposes a range of individual impressions that can be selected from when the notion is put to work:

> When we have found a resemblance among several objects, that often occur to us, we apply the same name to all of them, whatever differences we may observe in the degrees of their quantity and quality, and whatever other differences may appear among them. After we have acquired a custom of this kind, the hearing of that name revives the idea of one of these objects, and makes the imagination conceive it with all its particular circumstances and propositions. [...] They are not really and in fact present to the mind, but only in power; nor do we draw them all out distinctly in the imagination, but keep ourselves in a readiness to survey any of them, as we may be prompted by a present design or necessity. (Hume, 1978, p. 20)

From this two ingredients, vital to my understanding of tropes, can be extracted. First, both the existence of other tropes (in Hume's case impressions) and universals (in Hume's case abstract notions) are irrelevant to the individual natures of the tropes. A trope is enough for its own nature. Second, once tropes are introduced, universals (abstract notions) can plausibly be accounted for by some kind of bundle theory. Two other ingredients are provoked by the following passage from Hume:

> Upon this head we may observe, that all sensations are felt by the mind, such as they really are, and that when we doubt, whether they present themselves as distinct objects, or as mere impressions, the difficulty is not concerning their nature, but concerning their relations and situation. (Hume, 1978, p. 189)

Whether or not a trope (impression) is an object in its own right or not depends on whether it is object-related to further tropes or not. This implies two things. Just as tropes have a property-aspect, they have an object-aspect. So, first, tropes can be objects in themselves. Second, objects are either single tropes or some kind of bundles of tropes.

Tropes are thus bundled in different ways. They build individuals and they build kinds of properties. If we assume that there is a trope matching the linguistic expression "the brightness of the sun", this trope is probably assumed by us to be bundled in at least two ways – as a property of an individual, the sun, and as an instance of a kind of property, brightness (in general). So far trope Bundle-trope theorists, such as D. C. Williams and John Bacon (Bacon, 1995), would agree. Reading Bacon however makes one unclear about whether trope theory in the end differs much from nominalism. This is because in his writings the notion of property is always tied to the notion of universal, or kind of property. But the trope itself is the property. It is seldom correct to think of a trope as a universal since universals in many cases are shared by non-overlapping (complete) individuals. Neither is it in the majority of cases correct to think of a trope as a (complete) individual itself. Individuals typically consist of several tropes.

1.2 Tropes and relations

Not only properties but also relations, i.e. polyadic properties, are tropes. Not only *a's redness* but also *a's causing b* is a trope. By analogy, if the tropes *a's causing b* and *b's causing a* are different, as they mostly are, this difference resides in the tropes themselves. It isn't affected by whether and how these tropes are bundled.

Furthermore, as Ramsey (Ramsey, 1954b, p. 123) noticed, our interests are primarily focused on particulars and kinds of properties. Talk about *a's causing b* and *b's causing a* is naturally interpreted as talk about relations between the two individuals, a and b. In a trope context one is often led to regard such talk as if a two-term *relation trope* had the individuals a and b as its relata. Within the Bundle-trope framework I advocate any such talk "aRb" is interpreted instead as being about the intersection between the bundles constituting a, b, and R. If there is a trope (aRb) in this intersection, such talk is substantial. Despite appearances, *a's causing b* is then not a two-term relation between the bundles a and b. It is a trope that is bundled in at least three ways: as a, b, and as causing.

Though I will continue to use the term "relation trope", it has to be noted that when construed in the above way it is understood in a wider sense than the ordinary conception of "relation" permits. This is especially valuable for matters of causation. We tend to think of causation as a two-term relation between concrete relata (objects, events, facts, or tropes). But many instances of causation do not fit this pattern. There is also prevention, omission, etc. (Mellor, 1995, Persson, 2002). And, of course, in these cases causation is not a relation between absences. There is a more fundamental conception of causation in terms of my "relation tropes".

The reason why this intersection approach to polyadic universals is seldom recognised by trope theorists – even though it is the standard way they interpret one-term universals – is that the same approach can be disproved of within nominalism. A Bundle-nominalistic version of the intersection approach cannot account for asymmetric relations. The intersection between the three bundles a, b, and R does not help to distinguish aRb from bRa. This is of course bad news for the nominalist. Within this trope theory such differences are to be found in the nature of the tropes, rather than in any bundles or intersections. This is why the intersection strategy is more promising for the trope theorist who already has the properties and only needs the intersection of bundles to account for the idea of instantiating an n-term *universal.*

2. A Trope Interpretation of Laws

I will now apply the idea of how two-term universals are instantiated on laws and their properties. Let us assume that one typical kind of law is captured by the previous formulation "Objects with property P affect objects with property Q". According to the intersection view, an instance of this law, where a is P and b is Q, can be pictured as follows:

Both Relata 1) an intersection between the bundle a and the bundle P consisting of at least one trope a's being P; 2) an intersection between the bundle b and the bundle Q consisting of at least one trope b's being Q; 3) an intersection between the bundles a, b, and the bundle consisting of instances of the law in question.

What one has in every such case are three tropes, each figuring in a characteristic intersection.

2.1 The instance view of laws

Sometimes it is thought that a law is nothing but the collection of its instances, and the instances one has in mind are often the positive ones above, where both the relata of the law exist. If this view is satisfactory the law is exhausted by the situations where configurations meeting 1-3 of Both Relata are fulfilled. This extension of the intersection approach to relations then immediately yields a full account of laws.

However, a general worry with the instance view of laws is that most laws, by the use of the universal quantifier, contain more than these instances provide. The instances meeting 1-3 seem not to be enough to make true a universally quantified expression. Besides the instances the extra fact that these instances are all instances there are seems called for.

But while this might be sound as a critique of an instance-view regarding universally quantified expressions, it should not be presupposed that universally quantified expressions optimally capture the meaning of law formulations. First, it is not clear that they capture what exists in the world in a unique way (A). Secondly, it is not clear whether what they add has more to do with the role laws have for us than what laws are, ontologically speaking (B). (B) is part of Ramsey's later standpoint.[2] If only (B), the instance approach can be correct as far as laws in the world are concerned. For that reason one does not even have to assign a truth value to general expressions. In these circumstances they might without consequence be looked upon as rules for judging. If also (A), one has to see what more than the positive instances laws require. The natural thing to add to the view is that a law holds also in cases where its relata do not occur. We do not limit ourselves then to the rather trivial interpretation about instances still to occur. Instances of the law that will eventually happen can easily be accomodated by an atemporal view of the instances. There is a far more interesting interpretation that must be dealt with. The law of gravity, we might suppose, is not restricted to the times and places where objects it successfully acts on occur, but holds at all times and all places, irrespective of whether objects are there to be related by it. If this is the motivation for insisting on the universal quantifier, it is obvious that there are some alternative routes leading approximately there. This is since the intersection approach also permits of the following two instantiations of a law:

Partial Relata 1) an intersection between the bundle a and the bundle P consisting of at least one trope *a's being P*; 2) an intersection between the bundle a, and the bundle consisting of instances of the law in question.

One can have instances where the law exists and one but not the other relata exists. And consequently also where no relata exists:

Only Mechanism 1) a trope belonging to the bundle consisting of instances of the law (M) in question.

As indicated above, besides giving a wider spectrum of instances of a law, PARTIAL RELATA and ONLY MECHANISM are important in that they allow for causation without relata. A basis for all such cases seems to be had within the present approach.

Together the three of BOTH RELATA, PARTIAL RELATA, and ONLY MECHANISM provide new possibilities to account for laws through its instances. We are not confined to the traditional solutions that Russell criticised.

---

[2]See Ramsey, 1954a and Sahlin, 1990.

3. Trope Laws, Kinds of Properties, and Fake Properties

The above machinery gives an enhanced picture of the connection between laws and properties, *when* the laws pick out genuine properties. The relevant properties of laws are the properties of the objects the law relates, as well as the relation or mechanism (M) itself. The instance approach to general expressions is boosted by this acknowledgement of mechanisms. "Relation tropes" fulfil many of the roles that we usually attribute to general and modal relations.

Can this trope view also be utilised in distinguishing between properties and fake properties? Kinds of properties, i.e. universals, are treated as bundles. A trope can be bundled in several ways. If we think of universals as one "dimension" in which bundles exist, and particulars or objects as another, it doesn't seem counter intuitive to admit of multiple bundling also within one and the same dimension. It is then possible to regard one of the bundles in each dimension as more fundamental than the others. Thus, if we exemplify with the universals-dimension, the trope view can provide the ground for a distinction between genuine and fake universals. "When does new science (via new laws) entail new properties?" Part of the answer can easily be given: "When the bundles figuring in the intersections mark genuine universals, as opposed to when they are fake." For simplicity, let us distinguish between meta-relations of two kinds in the trope $\rightarrow$ universals dimension: G- and F-relations. A bundle united by G-relations is a genuine universal while a bundle united by F-relations is a fake universal. Since already the trope is a property, relations between two tropes in the trope $\rightarrow$ universals dimension might to a higher or lower degree depend on these intrinsic differences and similarities. It is possible that some relations exclusively depend on the tropes themselves, while others heavily depend on our interests, ignorance, cultural values, etc. From the metaphysical perspective, G-relations have more to do with what follows from the tropes than F-relations have.

4. Familiar Guides to the Properties: Maxwell and Ramsey

The difficulty with this categorisation is of course that in science we often start at the concept/universal level by formulating various hypotheses and theories, and then we ask what properties there are. That is, we do not typically know the natures of the tropes sufficiently well to determine which meta-relations are F and which are G. It is therefore interesting to see what ontological implications various guides to the properties have, i.e. what extra characteristics of G-relations they rely on. Clerk Maxwell's request for multiple display and what could be called Ramsey's test for properties attract many philosophers, and there is also a third one which I will label the concreteness test. Each of these guides builds on an alleged specific extra characteristic of G-relations (or G-bundles) that helps us distinguish between G- and F-bundles. I will end this paper by examining these three suggestions.

4.1 Maxwell's guide

Maxwell's request is that unless the assumed property manifests itself in several ways it should only be thought of as a concept: "If a quantity is connected to other effects which are independently defined then it is a physical state; if not then it is a mere scientific concept" (quoted from Turner, 1955, p. 233). To use one of Maxwell's examples: The only way to determine the electric force at a point is by an actual test charge or a calculation from the distribution of charge (the electric force at a point is defined as the force which would act upon a unit charge if placed there). The velocity of the fluid at a point, on the other hand, can be calculated and tested for in many ways, for instance by placing, at this point, a small cork tied to a spring. According to Maxwell's test, the velocity at a point has the mark of a physical state while the electrical force at a point has not.

There is thus an ontological assumption involved: *every instance of a G-bundle has (the capacity of) the same kind of multiple display.* That every instance (= trope) needs this capacity can be seen in the following way. Let us construct the disjunctive concept, $P = (F$ or G or $H)$. If Maxwell's guide were put to work on this concept, its applicability could be tested for in at least three ways; one associated with F, one with G, and one with H. It would be absurd to think that this proved that the new property P was discovered. To go with Maxwell's test is to claim that it is either the case that the mechanisms function in at least two kinds of 1-3 scenarios (so that in addition to P, Q, $\mathbf{M}$ intersections, there are R, S, $\mathbf{M}$ intersections), or that the tropes in the $\mathbf{M}$-bundles have that capacity.

Why should we believe in Maxwell's guide? It is interesting to note that the test belongs to a family of closely resembling methodological tools within the philosophy of science. Is it really a coincidence that Maxwell's request resembles what Popper and Lakatos demanded of acceptable post hoc reasoning? There seems to be no *a priori* reason why guides to such different areas should be so similar. I guess a principle of caution motivates it, exactly as in the debate over ad hoc hypotheses.

In that sense it doesn't seem to build on anything special to the laws' properties. It is a recommendable guide, I think, if one takes it as giving a reason for believing that something really is a property, but seems less reliable if it is understood as a necessary condition for something to qualify as a property. A principle of caution like this threatens to shut out certain properties from consideration: causally inert ones, for instance. If other proofs existed multiple causal powers should not be required. In any case, it assumes a substantial addition to the ontological framework we started out with.

4.2 Ramsey's guide

Ramsey's test identifies the properties in nature with what is quantified over when all law statements are conjoined and all predicates in this conjunction are replaced with variables. This procedure gives an at least imaginary Ramsey sentence which says that 'there are in the world properties that occur in this and that way in laws of nature'.

One problem with this view is that both fake and genuine properties were supposed to be involved in these laws. Ramsey's test seems to imply that unless there is a hierarchy of laws where only the more fundamental turn up in Ramsey sentences, all concepts in true laws correspond to properties in the world. Another problem is that it seems to make properties dependent on our concepts, but in this particular framework that doesn't need to detain us – we are only interested in cases where concepts in fact exists. What about the first problem? I think it is not solvable, and side with D. H. Mellor (Mellor, 1995, p. 192): "For Ramsey's test to tell us what we want to know, we need a so-called *objectual* reading of Σ's second-order quantifiers: that is we must take them to range over universals, not actual predicates." This shift is radical since it replaces the notion of law (as something that occurs in a theory) with something that exists in the world, and so already relies on a view of laws of different ontological status to provide the properties.[3]

5. Introducing the Concreteness Test

There is at least one more guide to the properties. Its starting point is the observation that sometimes one only comes up with new terms for properties already included in the world. Depending on how our

[3] There is also the possibility that some phenomena are not governed by laws. The Ramsey test seems to build on the assumption that everything that can be discovered figures in laws. In many sciences laws are rare. Does that mean the concepts they employ cannot correspond to properties in the world? The Ramsey test seems to fit some sciences much better than others.

conceptual frameworks develop, the relations between such new concepts and old properties will of course differ. If one for instance assumes a Comtian or Duhemian view, where the drift from concrete towards more abstract scientific concepts is of fundamental importance, a *concreteness test* for properties seems promising.

Whilst trying to block the realist's expansion of the universe, Nancy Cartwright advocates one such position: "The properties [the laws] mention are often already there; the new concepts just give a more abstract name to them" (Cartwright, 1999, p. 36). Cartwright presents a view where most laws contain abstract concepts and only models are concrete enough for marking properties. The distinctions between models, laws and their corresponding theories is not of concern here. It is the concepts' varying positions in the concrete/abstract spectrum and their relations to which properties there are that is important.

Before going into details, the tension that appears to arise between the three guides could be indicated. Let us assume that a certain theory develops in such a way that at t_1 the theory contains concepts P, Q, and R, and two laws, [P-objects affect R-objects] and [Q-objects affect R-objects], involving those concepts; while at t_2 the more abstract concept W has replaced P and Q, and the law [W-objects affect R-objects] has replaced the former two laws. Since W seems to be a more abstract name for what was previously referred to by P and Q, the concreteness test would have P and Q as marking properties, while W's link to the properties would be *via* P and Q – It wouldn't mark a property of its own. Whether the Ramsey test would give that result would depend on whether W figured in further laws in which P and Q did not and/or vice versa. In the former case the two tests would contradict each other. Arguably, if P and Q had multiple display, W would also, so from Maxwell's point of view all three concepts could name properties, which would again contradict the other tests.

6. What is Concrete/Abstract?

The traditional two guides are easy enough to understand but has little to do with the specific ontological problem posed here, and the concreteness test needs clarification but is closely related to the relevant ontological issues. The rest of the paper will focus on various guises of the concrete/abstract distinction.

Is the difference between atomic predicates and logically complex ones, such as $P \wedge Q$ or $P \vee Q$, mirrored in their position in the concrete/abstract spectrum? To me it seems reasonable to think of the difference in that way, and at least for some actual cases this way of approaching the concrete-abstract spectrum seems intuitively right. Yellow might be the disjunction of yellow shades, as Hume might have wanted it. Yellow seems more abstract to us than any of its particular shades. It is even clearer that colour is more abstract than yellow; and possibly colour is the disjunction of all colours. Temperature might be thought of as the disjunction of all temperatures, and so on.

As will be seen below, even if this way of understanding the relation between concrete and abstract works for some cases, it does not work for all. In those cases logical complexity catch an important aspect of the distinction, a conreteness criterion would have strong support from considerations of metaphysical sparseness. If P is accepted at t_1, a conceptual transition to $P \vee Q$ at t_2, shouldn't result in an ontological expansion of the world from P to $P \vee Q$, contrary to what the formula new science $\rightarrow$ new properties suggests. The obvious reason is that such a move would create double counting of properties (Compare Maurin and Persson, 2001). From the existence of the old property P would follow the existence of property P and new property $P \vee Q$. In such circumstances that could be described as abstraction through increased logical complexity, the concreteness test seems reliable enough. Such cases seem indeed to conform to the description: new names for properties that are already there.

Sometimes, however, logical complexity works in the reverse direction. It is easy enough to imagine cases where concepts are divided into component-concepts. Scientific progress is often achieved in that way. Recollect the useful distinctions between arteries and nerves, between heart and brain; and the division between kinetic and potential energy. In the light of new discoveries and conceptual changes, certain assumed properties could suddenly be understood as complex – as disjunctions or conjunctions of properties maybe. Taken in isolation, the transition from energy to kinetic energy can hardly be understood as a move in the abstract $\rightarrow$ concrete direction. What should be more abstract: energy at t_1 or kinetic energy at t_2? But when we at t_2 look back at our earlier concept we see that, if it persists, it becomes abstract in the logical complexity sense given above.

If being a disjunction is all there is to the former concept at t_2, is this an instance of the formula new concepts $\rightarrow$ old properties? The formula doesn't seem to correctly describe what is going on. Properties are not added by abstraction from the concrete, we are engaged in a more analytical enterprise, refining our views on what properties there are or how they are composed. How the concreteness test deals with analytical situations is not entirely clear. As I see it there are only two options of which the first makes little sense.

First, one can rely on the *temporal condition.* If the concepts at t_1 and t_2 are associated, the ones at t_1 are more concrete. Hence only the earlier concepts mark properties. This reading fits Cartwright's formula new concepts but old properties best. But it seems at all plausible only if we already assume an exclusively Comtian or Duhemian view. It does not capture much of daily scientific practice. In particular it doesn't seem to go along with analytical phases of the scientific enterprise. A concreteness test modelled on this condition is implausible.

Second, one can use the process of abstraction only as an *exemplification of how the relations at hand may arise.* The double counting argument lets us decide that whenever we have a situation where three concepts are linked to one another as disjuncts to disjunction or conjuncts to conjunction, at least one of them does not exist as a property. In a Duhemian world it is probably the concepts developed earlier that have matching properties, while in the analytical phase it might as well be the later. But then for this to be a test, a non-temporal understanding of the terms "concrete" and "abstract" is needed.

7. Further Examples and a Bundle-Trope Characterisation

There are further reasons why the concrete/abstract distinction needs more work. Sometimes conceptual changes in a theory do not establish the kind of logical relationships we have examined. Even the concept of energy has more to it than being the disjunction of kinetic and potential energy. It operates on other levels of the theory as well. This is obvious for conceptual relations in ordinary language. To make this point, Cartwright uses the relations between working, washing dishes, writing a proposal, etc, i.e. concepts describing activities she was engaged in a particular day:

> *Work* has implications about leisure, labour, preference, value, and the like, that are not already there in the description of my activity as washing the dishes or negotiating with the dean. (Cartwright, 1999, p. 40)

Yet:

> *Working* is a more abstract description of the same activities I have already described when I say that I washed dishes, wrote a proposal, and bargained with the dean. (Cartwright, 1999, p. 40)

Let us take it for granted that Cartwright is right in these two observations. Then how is concrete/abstract to be understood in this context? Let us begin by formulating the example in terms of the Bundle-trope approach. That there is in fact no difference between the particular property, that in this case counts as washing dishes, and the property, that is also categorised as working, should mean that one and the same trope occurs in two bundles: the washing- and the working-bundle.

The concreteness test now says that we should go for washing dishes as being both a concept and a property, and treat working only as a concept. Why is the one bundle more genuine than the other?

The double counting argument provided a suitable ground before, will it work again? A complication has emerged. Before, although the concepts were differently positioned in the concrete/abstract spectrum, the properties were supposed to exist on the same level. But if concepts are of such different hierarchical levels as Cartwright claims is the case for work and washing dishes, then why cannot properties also exist on mirroring hierarchical levels? Perhaps kinetic energy would qualify as a first-order property and energy as a property of second-order. The argument from double counting could then not be applied on the level of properties. An instance of the property kinetic energy would not be an instance of the property energy, although of course the body that would instantiate kinetic energy would also instantiate energy. A less complex version of this difficulty emerges already when we compare kinetic energy with specific amounts of kinetic energy. It is what is nowadays again referred to as the determinate/determinable distinction. It makes already being a certain determinable (e.g. a temperature) a second-order property, namely a property of certain first-order properties (e.g. 100°C), the determinates (Maurin and Persson, 2001).

Couldn't a similar response to the double counting argument be used against "hierarchical promiscuity"? Only traditional arguments from ontological economy, which however much practised are difficult to give a solid basis. Different versions of Occam's razor are extensively used but their merits are merely presupposed. Yet, if one is happy with Occam's razor it speaks in favour of the concreteness test also on this slightly more complicated hierarchical level.

8. Abstract Concepts as Concepts that Need 'Fitting Out' in a Concrete Way

Cartwright's most ambitious attempt to argue in favour of a concreteness view consists in offering the way fables transform the abstract into the concrete as mimicking how models function in physics. What is abstract in the first case is the moral; what is, according to her, abstract in the second case is the scientific law. She borrows a theory about the relation between morals and fables from Lessing. Three different functions are assigned to the fable: 1) by making the content concrete (or rather intuitive as opposed to symbolic) it makes the moral possible to visualise, and so *testable*; 2) intuitive recognition has stronger motivational force than symbolic recognition; and 3) the abstract can only exist in the particular (Cartwright, 1999, p. 38).

Cartwright takes interest only in the last, ontological claim. That is a mistake since the first function is important too. If the process of making concepts more concrete, as when a fable is constructed to "fit out" a moral, results in descriptions that are easier to test and visualise, then epistemological considerations clearly speak in favour of the concreteness test. If Lessing is right, the concreteness test can be backed up by appeal to considerations of discernability. Such a claim would fit nicely with views such as van Fraassen's and many other empiricists. Even if we have abandoned the stubborn view that only what we are able to perceive exists, there is little reason to deny that what we perceive exists. From the point of view the concreteness guide offers, this function of fables is highly relevant, whether or not the more abstract theory is efficiently tested by the concrete consequences.

Of course, the bolder claim that the abstract can only exist in the particular must be more attractive to Cartwright. It is not that much discussed, and it would be suitable to have a clearly ontological argument at this point. According to Lessing and Cartwright, the more abstract description never applies unless the more concrete one does. Moreover, satisfying the concrete concept is what satisfying the abstract description consists in on that occasion. Consider:

> Moral: The weaker are always prey to the stronger.
> Fable: A marten eats the grouse; A fox throttles the marten; the tooth of the wolf, the fox.

Neither of the concepts: the weaker, the stronger, or being prey to, appears in the fable; and they are not needed to fix the factual picture. The marten is wily and quick; the grouse is slow and innocent. "That is what it is for the grouse to be weaker than the marten" (Cartwright, 1999, p. 41). The wolf is bigger and has sharper teeth than the fox, and that is what it is for him to be stronger than the fox.

According to this picture, abstract concepts are satisfied by the existence of properties corresponding to qualitatively different concepts. A merit of the fables and morals comparison is that we are presented with a greater conceptual difference between the two concepts. The fable, like the model, constructs an isolated enough context where the concrete properties will interact in a way that makes descriptions containing the associated abstract concepts be true of the situation. To "fit out" the abstract concepts in this way is how Cartwright thinks more abstract concepts relate to the world.

Is she right? I am not convinced. It could for instance be that more concrete properties in specific circumstances caused more abstract properties to exist (or provided a basis for them in some other way). Since such a relation on the level of the properties would be a context-dependent relation, showing that context-dependent relations exist on the conceptual level support that possibility rather than the conclusion that there are no abstract properties of that kind. As far as I can see, the context-dependence claim in 3 supports only another of Cartwright's theses, namely that the scope and truth of law statements is a tricky business that makes attractive the view that laws only hold in those contexts where we can construct a model that actually conforms to what the law says should happen. This is an interesting enough claim, but it has no bearing on what properties exist in the world. Existing in a particular context would obviously be enough for existing in the world.

A further idiosyncrasy can be seen if one considers the point where concepts become so concrete that one normally would have no epistemic qualms about accepting them as marking properties. If the concreteness test is still applicable, it forces us to order such concepts – one more abstract than the other. Only the more concrete should be accepted. The question of when and why fitting out is a reliable characteristic of G-bundles remains open. When it comes to the concreteness test one should perhaps rest content with the empiricist conviction that what can be perceived should count as a property.

9. Conclusion

We are good at discussing law statements of different epistemic status, and to describe logical relationships between different law statements. But I feel that contemporary discussion often suffers from its difficulty to formulate questions concerning laws of different *ontological* status. This paper has presented a framework for distinguishing between properties and fake properties that seems to provide better tools for such inquiries. It also helps us transcend two other barriers that have been limiting in our attempts to understand the laws of nature. The first problem is the seemingly inevitable use of universally quantified expressions when thinking about laws. The second is our incapacity to abandon our understanding of causation and other instances of laws in terms of relations between two or more relata. Both these barriers frequently lead us wrong, and in the text I have indicated why I think so.

This paper has also put the framework to use in discussing three suggested tests for properties, suggested by Maxwell, Ramsey, and Cartwright. None of these tests is good as it stands. This points in favour of a view where we, rather than favouring one particular test, are looking for methodologically stable decisions, i.e. decisions where several tests come to the same conclusion.

References

Bacon, J. (1995). *Universals and Property Instances.* Blackwell, London.

Cartwright, N. (1999). *The Dappled World.* Cambridge University Press, Cambridge.

Hume, D. (1978). *A Treatise of Human Nature.* Clarendon Press, Oxford.

Kistler, M. (1999). *Causalité et lois de la nature.* Librairie philosophique J. Vrin, Paris.

Maurin, A.-S. and Persson, J. (2001). Realistic metaphysics: An interview with D. H. Mellor. *Theoria*, LXVII:96–113.

Mellor, D. H. (1995). *The Facts of Causation.* Routledge, London.

Persson, J. (2002). Cause, effect, and fake causation. *Synthese*, 131:128–143.

Ramsey, F. P. (1954a). General propositions and causality. In Braithwaite, R. B., editor, *The Foundations of Mathematics*, pages 237–255, London. Routledge and Kegan Paul.

Ramsey, F. P. (1954b). Universals. In Braithwaite, R. B., editor, *The Foundations of Mathematics*, pages 112–134, London. Routledge and Kegan Paul.

Sahlin, N.-E. (1990). *The Philosophy of F. P. Ramsey.* Cambridge University Press, Cambridge.

Turner, J. (1955). Maxwell on the method of physical analogy. *The British Journal for the Philosophy of Science*, 6:226–238.

Williams, D. C. (1953). On the elements of being. *The review of metaphysics*, VII:3–18, 171–192.

LAWS OF NATURE VERSUS SYSTEM LAWS

Gerhard Schurz

1. Introduction

What distinguishes accidental generalizations such as

1 All apples in this basket are red

from truly lawlike generalizations? – such as

2 All masses attract each other – or

3 All ravens are black

Starting with the famous papers of Goodman (Goodman, 1946), Carnap (Carnap, 1947) and Hempel/Oppenheim (Hempel and Oppenheim, 1948), this question has been discussed for decades. By way of introduction I mention five important problems which are involved in the topic of lawlikeness:

1.1 Universality or necessity?

A straightforward suggestion which separates the accidental generalization (1) from the lawlike generalizations (2,3) is the condition of *spatiotemporal universality*: lawlike generalizations must make nontrivial assertions for every spacetime region, not only for "this basket" as in (1) (cf. Earman, 1978, Schurz, 1983, ch. VI.1). A stronger but related suggestion of Carnap and Hempel is the so-called *Maxwell-condition* which says that lawlike generalizations must not refer to particular individuals. The problem with these proposals for lawlikeness criteria is that, in spite of their plausibility, they seem to miss the point. There are lots of purely universal but completely accidental generalizations. To take an example due to Reichenbach,

4 All solid spheres of gold have a diameter of less than one mile

J. Faye, P. Needham, U. Scheffler and M. Urchs, (eds.), Nature's Principles, 255–268.

is such a completely universal but nevertheless accidental generalization.

So, on the one hand, the condition of spatiotemporal universality or the related Maxwell condition is too weak for lawlikeness. On the other hand, there are also arguments to the effect that these conditions are too strong. For example, they are probably too strong in the context of general relativity theory, as Earman has pointed out in (Earman, 1978). Or, they are clearly too strong for so-called derived laws, to which I will turn in section 1.4

Philosophers like Armstrong (Armstrong, 1983) conclude from these problems that the crucial criterion for lawlikeness is not universality but *necessity.* For example, a sphere of gold with a diameter of more than a mile is physically *possible*; hence (4) does not express a physical necessity – in contrast to a sphere of radioactive uranium with such a diameter, which cannot exist because it its mass by far exceeds the critical mass for an atomic explosion. However, the assumption of necessities of nature is a metaphysical stipulation, and it is not clear which insights it should bring us (apart from the intrinsic problems of Armstrong's "necessitation-between-universals"-account; cf. van Fraassen, 1989, 94ff). Moreover, there exist many laws which do not satisfy Armstrong's strong intuitions of necessity and yet are intuitively considered as laws. For example, our example (3) that all ravens are black is not lawlike according to Armstrong (cf. Armstrong, 1983, 18). But here Armstrong's intuition is in conflict with a commonly accepted indication of the lawlikeness of a generalization, namely that it supports *counterfactual claims.* Generalization (3) supports the counterfactual claim that if this bird were a raven, it were black – contrary to Armstrong's view.

1.2 Laws of nature as rules of language?

Another account which turns laws of nature into necessarily true or even normatively valid statements is the view that they express linguistic conventions, or rules of language. This view has been supported by the conventionalists, with Henri Poincaré as their famous representative, and a variant of this view is defended by Jan Faye in his paper in this volume. The main problem which I see with this account is that laws of nature such as Newton's second force law, force = mass times acceleration, if joined with other principles, do obviously produce empirical content by which they can be empirically confirmed or weakened (modulo the famous "Duhem-holism of theory-falsification"). For example, the vector version of Newton's second force law is no longer valid in the special theory of relativity: if a force f is acting on a massive body x with velocity v, then the resulting acceleration of x is no longer parallel to f in the case where f and v are not parallel and the inertial system is different from the eigen-system of x (cf. French, 1968, ch. 7.4; this is a comparatively mild departure of classical physics as compared to the general theory of relativity, or to quantum mechanics). I take it to be obvious that Einstein, in his special theory of relativity, has not just invented a more comfortable rule for physicist's linguistic behaviour – rather, Einstein has found something *new* about the physical world. I conclude that laws of nature are not rules of language but candidates of better or worse approximation to the truth.

1.3 Laws of nature as the premises of the best unification

An approach which is, at least in my view, a more promising one is the so-called Mill-Ramsey-Lewis account (cf. Earman, 1986, 88) according to which the lawlike generalizations are those which are derivable from those theories which achieve the best unification of the set of all true statements. This approach is right insofar what we usually regard as laws of nature are core principles of theories with a high unification power. So, unification is probably a necessary condition for laws of nature. But it does not seem to be also a sufficient condition. For observe that accidental generalizations are usually no derivable from theories, not even from theories plus simple boundary conditions. Hence, why should an accidental generalization such as (1) or (4) not figure as a premise in the best unification of the set of all true statements? If this unification should deductively cover *all* true statements, and hence also (1) and (4), then it seems that it would have to contain (1) and (4) among its premises. If this unification covers only a 'great amount' of these true statements, then the ques-tion is, *which* amount should it cover, and *how* should one balance the *stringency* versus the *breadth* of a unification, for one goes on the cost of the other. In other words, there exists no unique, non-arbitrary measure for unification – the best what can be done is a partial ordering of belief systems with respect to their unification (cf. Schurz and Lambert, 1994). This means that whether (1) and (4) will come out as laws according to the unification approach may depend on arbitrary features of the unification measure.

1.4 Fundamental vs. derived laws and the gradual transition problem

Connected with the unification problem is another one, that of the status of generalizations which are derived from laws of nature. For example,

5 Galileo's law of the free fall (all bodies have the same free fall acceleration of approximately $10m/\sec^2$)

refers to the earth; and yet this generalization is lawlike and supports counterfactual claims, such as if you were to jump out of the window you would fall to the ground. For this reason, Hempel suggested in (Hempel and Oppenheim, 1948) that only fundamental laws must be universal, while derived laws may contain spacetime restrictions or individual constants. But in (Nagel, 1961) Ernest Nagel has pointed out that the following problem is lurking behind this suggestion: usually, derivative laws are not derived from fundamental laws alone but from fundamental laws plus certain factual premises, so-called *boundary conditions*. For example, Galileo' laws is derivable from classical physics and boundary conditions about mass and diameter of the earth. But the same is true for accidental generalizations such as

6 All screws in Smith's cars are rusty

Generalization (6) is derivable from the fundamental law "all iron exposed to oxygen rusts" plus suitable factual premises about Smith's car which imply that this car has very often been exposed to rain. A closer look at this example makes us even more uncertain about its nature, for given the circumstance of continuous exposure to rain it seems to be counterfactually true that if this screw were one of Smith's car, it were rusty.

Hence, some *amount* of lawlikeness, so to speak, is even contained in example (6). It seems that opening the category of derived laws forces us into a *gradualization of lawlikeness* – lawlikeness of generalizations becomes a more-or-less notion. An alternative suggestion would be to count only fundamental laws as truly lawlike (as has been proposed by Flichman, 1995) – but again, this requirement would somehow miss the point. On the one hand, this proposal would rule out too many intuitively lawlike generalizations, like Galileo's law, Kepler's laws, Mendel's law of genetics, etc. On the other hand, the requirement of *logical fundamentality* is itself not essentially connected with lawlikeness, because – as we have pointed out already in 1.3 – there are lots of special and even accidental generalizations which are the result of indeterministic processes and, thus, are not derivative but fundamental laws, concerning their logical status. For example, our raven-law (3) or the gold-generalization (4) are fundamental in this sense.

Summarizing, the criteria discussed so far seem to miss the explicandum of "lawlikeness". In the next problem the situation gets even worse: we are no longer sure that the explicandum exists.

1.5 Ceteris paribus laws – are there any laws at all?

The lawlike generalizations discussed above are formulated as if they were strictly true, but a closer look at the examples shows that none of them is really exceptionless. For example, masses will not attract each other if they are, in addition, electrically charged and repel each other. It seems that the law of gravitation holds only if further 'disturbing' factors are absent; which means that this law – as well as all other examples – must be understood as a so-called *ceteris paribus* law, in short a CP-law. This was Nancy Cartwright's argument in (Cartwright, 1983) from which she concluded that all laws of physics lie. Van Fraassen (van Fraassen, 1989) or Giere (Giere, 1999) have concluded from such problems that *there are no laws in nature.*

The sort of CP-laws which were central in the philosophical debate are what I call *exclusive CP-laws.* An exclusive CP-clause states that except for the causal factors mentioned in the law's antecedent, further interfering factors – so called disturbing factors – are *absent.* Here are some kinds of 'loose' laws, i.e. laws with exceptions, which have been reconstructed as exclusive CP-laws:

7 CP, planets move on elliptic orbits.

8 CP/Normally, birds can fly.

9 CP/Normally, people act goal-oriented.

A crucial distinction is that between *definite* and *indefinite* CP-laws. In definite CP-laws the CP-clause can be replaced by a (finite) *list of all possible disturbing factors* which are excluded in the antecedent of the CP-law. Such a transformation is called a *strict completion* of a CP-law. Definite CP-laws are harmless because they are eliminable in principle. The really crucial case are indefinite CP-laws. Here the number of possible exceptions is unknown and/or potentially infinite so that we cannot exhaustively describe them. Various philosophers have argued that indefinite CP-laws are *vacuous tautologies,* while other philosophers have tried to develop non-vacuous reconstructions of CP-laws. In Schurz,

2001b, Schurz, 2002 I have tried to show that even very elaborate reconstructions of CP-laws such as that of Pietroski and Rey (Pietroski and Rey, 1995) are almost vacuous; hence they are pseudo-laws. I argue that there are only two kinds of 'loose' laws which do have empirical content and are important in science, namely on the one hand *theoretically definite CP-laws* which occur in physics, where the disturbing factors are under theoretical control, and on the other hand so-called *normic laws* of the form "*A*s are normally *B*s" (cf. Scriven, 1959). On this reason, I have put "Normally" as an alternative reading of "CP" in the examples 8 and 9, but not in example 7. I will say more about normic laws soon. First, I want to summarize my introduction.

I have tried to show that the question of lawlikeness leads into a host of rather heterogeneous problems, and that the proposed criteria somehow miss the topic: they are of logical and epistemological, but not of substantial importance for our understanding of the nature of scientific laws. In the following I will suggest a classification of laws which intends to raise more substantial points. Is not logical or epistemological but ontological in nature (cf. also Josef Schurz: Schurz J., 1990).

2. Laws of Nature versus System Laws

Laws of nature are those fundamental laws of physics which hold everywhere in the universe, or in other words, which are not restricted to special entities. There are only a few of them. In classical physics, the *total force* law $F(x,t) = m(x).d^2s(x,t)/dt^2$ is a law of nature. It is a differential equation in which $F(x,t)$ figures as a *variable* function denoting the sum of *all* forces acting at time t on particle x without saying what these forces are. An analogue exists in quantum mechanics: the general Schroedinger equation with a *variable* operator for the total potential energy (cf. Anderson, 1971, pp. 141–143, 246). Another kind of laws of nature are *special force laws*, e.g. the classical laws for gravitational force or electric force – provided they are understood as laws about *abstract* component forces, or 'capacities' in the sense of Cartwright (Cartwright, 1989, pp. 183ff). Laws of nature are *strictly* true, *without* any ceteris paribus clause – but at the cost of not *per se* being *applicable* to *real* systems, because they do not specify *which* forces are active in the system under consideration.

System laws, in contrast, do not apply to the whole universe but speak about concrete systems x of a certain kind S in a certain time interval Δt, with a *specification* of all forces acting within or upon the system x in the time interval Δt. The specification of forces is the task of so-called *boundary conditions*. Examples of system laws of classical physics are

Kepler's laws of elliptic planetary orbits, or the laws describing a gravitational pendulum, a viscous fluid, etcetera – almost all laws in physics or chemistry are system laws. The boundary conditions of Kepler planetary systems are that the masses of the planets are small as compared to the sun and that the only non-negligible force acting upon them is the sun's centripetal force (hence, interplanetary forces and influences coming from the outside the solar system are neglectibly small).

The next important distinction is that between *theoretical* and *phenomenological* system laws. Theoretical system laws state the *special* differential equations for the kind of system under consideration, with a concrete specification of boundary conditions (forces). Phenomenological system laws describe the temporal behaviour of the system in an empirical or pre-theoretical vocabulary, and dependent on given initial conditions. For *simple* systems like the planets, the theoretical system laws are literally derivable from laws of nature and boundary conditions, by inserting '*ideal*' boundary conditions into the total force law. By solving the special differential equations of theoretical system laws, one may derive the phenomenological system laws which approximately describe the system's trajectories, i.e., its time-dependent development. In the planet example, these are Kepler's elliptic trajectory laws. But phenomenological system laws may also be obtained by purely empirical-inductive means without any theoretical derivation.

Let me now make three important points about this distinction. *The first point* concerns ceteris paribus clauses. It is system laws where CP-clauses are needed, not laws of nature. Laws of nature are silent about which forces are realized. In contrast, system laws specify which forces are realized in a particular kind of system, or at least, they depend on such a specification. It is here where one needs an exclusive list: this and this are the non-neglectible forces and *nothing else*. Therefore, system laws need a kind of CP-clause, which may be either explicit and exclusive or, as we shall see, implicit and normic. This observation may shed some light on Nancy Cartwright's change of position. In (Cartwright, 1983, p. 57f) she argues that the law of gravitation fails for electrically charged bodies. Here I think she treats this law wrongly as a system law. In (Cartwright, 1989, p. 192f) Cartwright changes her view: there she considers special force laws as abstract laws about capacities which hold *without* a ceteris paribus clause, hence she treats them as laws of nature in my sense.

The second point explains why the difference between laws of nature and system laws does *not* coincide with the *logical* distinction between universal versus spatio-temporally restricted laws. Assume that we would be able to state a so-called *strictly completed* version of Kepler's system law by relativizing the Kepler equations (abbreviated as KepEq(x,t)) to an antecedent condition (KepSys(x,t)) which specifies all boundary conditions of Keplerian planetary systems. The resulting universal implication

$$\forall x \forall t(\text{KepSys}(x,t) \rightarrow \text{KepEq}(x,t))$$

would be universally true and derivable from laws of nature *alone* without employing additional factual conditions – because all of them have now been shifted into the antecedent KepSys(x,t). But still, this implication would not count as a law of nature, but as a system law, because its antecedent is restricted to systems of certain kinds, while laws of nature are *not* subject to such an antecedent restriction.

It is a *third important point* that my distinction does not coincide with that between fundamental versus derived laws. On the one hand, there are not only fundamental laws of nature, there are also derived ones, namely derived from fundamental laws alone *without* inserting boundary conditions about special systems into them. This is related to a point made by Joseph (Joseph, 1980, p. 789) who – after finding that most laws of physics are CP-laws – recognizes that there are also some laws of physics which are literally true, such as the *conservation laws*. In our framework this has the following explanation: the conservation laws for energy and momentum are *directly* obtained from laws of nature by integrating the total force law over space or time, respectively, *without* any insertion of special boundary conditions – so these laws are derived laws of nature and they are true without CP-clauses.

On the other hand, most system laws, even many system laws of physics, are not derived laws but are themselves fundamental. Cartwright (cf. Cartwright, 1983, pp. 104f, 113f) and others have repeatedly demonstrated that for physical systems of moderate complexity, the theoretical system laws are not literally derivable from laws of nature and boundary conditions, nor are the phenomenological system laws literally derivable from the theoretical system laws. In the case of non-physical system laws such as that birds normally can fly, the attempt to derive them from laws of nature and boundary condition is usually completely hopeless. Phenomenological system laws of this kind are usually obtained by purely empirical-inductive means.

3. Closed (Isolated) versus Open (Self-regulatory) Systems

The second important distinction in my classification is the *system-theoretic* distinction between closed or isolated versus open systems. In closed systems, there is *no* exchange between system and environment;

in isolated systems, there is exchange of heat-energy, but no exchange of matter. Only in open systems is there a continuous exchange of both matter and energy between system and environment (cf. Bertalanffy, 1979, Rapaport, 1986, Schurz J., 1990).

The systems studied by physics or chemistry are, at least traditionally, closed or isolated systems. An example of closed system laws are, again, Kepler's laws of elliptic planetary orbits. Physical system laws are typically expressed as *strictly* universal generalizations. But as I have already explained, they must be furnished by ceteris paribus clauses which explicitly require that the system is closed, i.e., that no further and possibly disturbing factors are present. Thus my *first thesis* about system laws is that system laws of physics are, at least typically, exclusive CP-laws which must be theoretically definite in order to avoid emptiness of their content.

In contrast, all 'higher' sciences, from biology 'upwards' to social sciences and humanities, are concerned with open systems, more specifically with 'living' systems or with their cultural and technical products. Very generally, *systems* are physical ensembles composed of parts which preserve a relatively strict *identity* in time, by which they delimit themselves from their (significantly larger) environment (Rapaport, 1986, pp. 29ff). For closed system this preservation of identity follows from their isolation which, in turn, is a matter of *postulate*: that our planetary system is stable is a *frozen accident* of cosmic evolution; should it be devastated by a gigantic swarm of meteorites at some time, then it stays so forever thereafter and will *not* regenerate. But how can we explain the relatively strict identity of open systems, which are permanently subject to possibly destructive influences from the environment? The explanation lies in the fact that all open 'living' systems have the characteristic capacity of *self-regulation*. According to the framework of *cybernetics* (cf. Ashby, 1961), the identity of self-regulatory systems is abstractly governed by certain *norm states*, which the system constantly tries to approximate by its *real states*. It does this by certain *subsystems* (in biology: organs) which perform certain *regulatory mechanisms* (in biology: functions) which *compensate* for *disturbing influences* of the environment by producing counteracting processes. Thus, my second thesis is that *normic laws are the phenomenological laws of self-regulatory systems.*

But – why are self-regulatory systems omnipresent in our world? Where do their prototypical norm states *come from*? Why do their self-regulatory mechanisms normally work properly? The answer is: by *E*volution (with capital "E") in a generalized 'Darwinian' sense of evolution by natural or cultural *selection*. I call self-regulatory systems which have evolved by Evolution *evolutionary systems*. Their prototypical (norm) states are those states which are essential for their survival. Their self-regulatory mechanisms are those capacities which have been gradually selected in Evolution, according to their contribution to reproductive success. Due to their *limited* compensatory power, dysfunctions may occur, hence their normic behaviour may have various *exceptions*. Yet it must be the case that evolutionary systems are in their prototypical norm states in the high *statistical majority* of cases and time-points. For otherwise (with high probability), they would not have *survived* in Evolution. In this way, evolution theory explains not only why the phenomenological behaviour of evolutionary systems obeys *normic* laws – it explains also why this peculiar connection between *prototypical* and *statistical* normality exists at all. Birds, for example, can normally fly. Of course it is possible that due to a catastrophic event, all birds loose their flying ability, but then (with high probability), they will become extinct after a short period of Evolution. For similar reasons, electric installations normally work, for they are constructed in that way, and if this were not so, they could not survive in the economic market.

My third thesis, therefore, says almost all self-regulatory systems have evolved by Evolution, which implies by means of my second thesis that normic laws are the *phenomenological laws of evolutionary systems*. Thereby, prototypical normality and statistical normality are connected by the law of evolutionary selection.

This was a rather sketchy presentation of my evolution-theoretic foundation of normic laws which I have suggested and elaborated in Schurz, 2001a. Summarizing, I arrive at the following classificatory schema of laws which is represented on the next page.

It follows that the system laws of physical sciences and that of nonphysical sciences are rather different – exclusive CP-laws on the one hand, normic laws on the other. This difference is explained by the difference between closed (or isolated) and open self-regulatory systems. For closed (or isolated) systems, a detailed specification of *all* forces ('and nothing else') is needed, whence we must describe their behavior by means of exclusive CP-laws. For open self-regulatory systems, such a specification is neither possible nor necessary. It suffices to assume that the disturbing influences, whatever they may be, are within the 'manageable range' of the system's compensatory power. It is also usually impossible to give an exact theoretical prediction of this 'manageable range'. But evolution-theoretic considerations tell us that *normally* the external influences will be within this manageable range. This explains the *normic* character of the CP-laws of open self-regulatory systems.

To avoid misunderstandings, I do not claim that the borderline between closed or isolated physical systems and open self-regulatory sys-

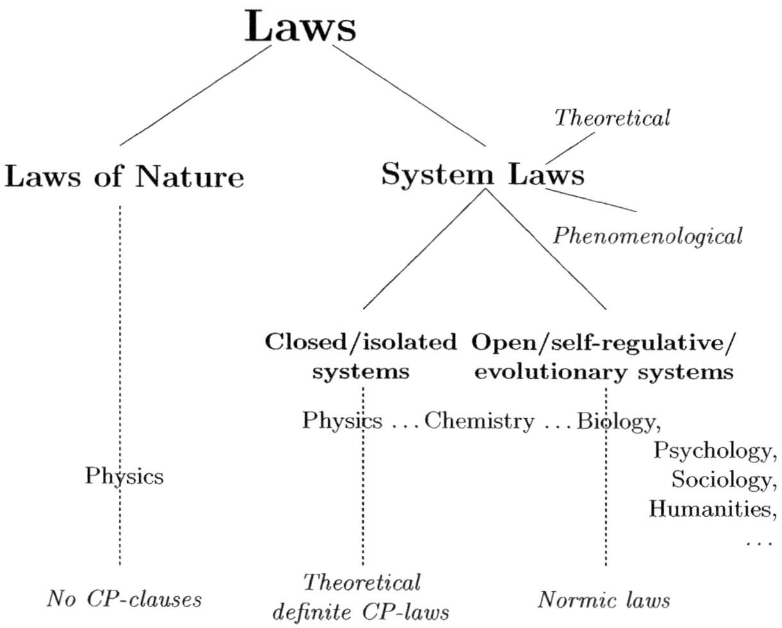

tems is strict – there are also transition cases between them, e.g. in areas like chemistry or geology. It is also clear that, in principle, one may always try to describe open systems as parts of larger closed physical systems (ultimatively the universe) – but in most real examples this would be a theoretically hopeless enterprise. However, there are systems which can be fruitfully described both as closed systems of physics and as parts of open evolutionary systems – namely *technical systems*. Consider the systems of electricity which surround us every day. We may consider this automatic dish washer together with its electric circuit as an ideally closed physical system. From this perspective, there are thousands of possible disturbing factors which may prevent our dishes from being cleaned, and amazed we may ask ourselves why all these electric systems can be so cheap and yet work so well. Alternatively, we may consider them as part of an evolutionary system – the economic system of production and distribution of electric products. This perspective does not give us detailed knowledge of the physical mechanisms underlying dish washers, but it gives us an evolution-theoretic explanation of their amazing optimisation of cheapness and functionality.

4. Conclusions

Let me summarize how the suggested classification of laws may shed new light on some old problems:

4.1. Those numerous people who consider CP-laws or normic laws as pseudo-laws are right if one reconstructs them as indefinite exclusive CP-laws. But I have argued that this is the wrong reconstruction; in physics they are theoretically definite exclusive CP-laws, in evolutionary sciences they are normic laws with statistical majority implications.

4.2. Those people (such as Schiffer, 1991) who argue that there are no genuine laws outside of the area of physics are right in that there are no non-physical laws of nature; but all sciences have their own system laws. That their system laws are indeed lawlike can be seen from the fact that they support counterfactual claims.

4.3. On the other hand, system laws also contain a considerable portion of accidentality; because they result from laws of nature *plus* factual boundary conditions. In particular, normic system laws are the result of laws of nature plus evolutionary processes. This explains the gradual transition between lawlike generalizations and accidental generalizations which we have noticed in the introduction – this graduality is an unavoidable phenomenon of system laws, but not one of laws of nature.

4.4. Laws of nature do not refer to any *system description*, not even to our entire universe conceived as a system. For example, that our universe is composed of matter is not a law of nature but a system law; here the entire universe is the system under consideration. Laws of nature are not only intended to speak about our universe but to speak about other possible universes. Armstrong's necessity account of lawlikeness may apply to laws of nature, but certainly not to system laws. There is also an important practical point here: laws of nature cannot possibly be changed by human powers. System laws, on the other hand, may be changed, provided they are not formulated in a strictly completed manner. For example, that all ravens are black, or that planets move in elliptic orbits, could in principle be changed by human powers.

4.5. Similar considerations apply to the condition of universality. Universality holds for laws of nature – for them this condition is too weak, as we have seen – but universality is too strong for system laws, because they depend on special boundary conditions which may be different in different spatiotemporal regions of the universe.

4.6. Our distinctions may shed some new light on other notorious questions of philosophy of physics. For example, consider the debate whether physical laws are complete and what this could mean. Of course, it can only mean whether physics is complete with respect to its laws of nature, because there are myriads of system laws and most of them are unknown to us.

Acknowledgements. For valuable discussions I am indebted to virtually all participants of the symposium. In particular, I want to thank Geert Keil and Paul Needham for various helpful remarks.

References

Anderson, E. (1971). *Modern Physics and Quantum Mechanics.* W. B. Saunders, Philadelphia.

Armstrong, D. (1983). *What is a Law of Nature?* Cambridge University Press, Cambridge.

Ashby, W. R. (1961). *An Introduction to Cybernetics.* Chapman & Hall, London.

Bertalanffy, L. (1979). *General System Theory.* New York, 6th edition.

Carnap, R. (1947). On the application of inductive logic. *Philosophy and Phenomenological Research*, 8:133–147.

Cartwright, N. (1983). *How the Law of Physics Lie.* Clarendon, Oxford.

Cartwright, N. (1989). *Nature's Capacities and their Measurement.* Clarendon, Oxford.

Earman, J. (1978). The universality of laws. *Philosophy of Science*, 45:173–181.

Earman, J. (1986). *A Primer on Determinism.* Reidel, Dordrecht.

Flichman, E. (1995). Hard and soft accidental uniformities. *Philosophy of Science*, 62:31–43.

French, A. P. (1968). *Special Relativity.* MIT, Cambridge/Mass.

Giere, R. (1999). *Science without Laws.* University of Chicago Press.

Goodman, N. (1946). A query on confirmation. *Journal of Philosophy*, 44:383–385.

Hempel, C. G. (1965). *Aspects of Scientific Explanation (and Other Essays).* The Free Press, New York.

Hempel, C. G. and Oppenheim, P. (1948). Studies in the logic of explanation. *Philosophy of Science*, 15:135–175. (Reprinted in Hempel, 1965: 245–290).

Joseph, G. (1980). The many sciences and the one world. *Journal of Philosophy*, 77:773–791.

Nagel, E. (1961). *The Structure of Science.* Routledge & Kegan Paul, London.

Pietroski, P. and Rey, G. (1995). When other thing's aren't equal: Saving ceteris paribus laws from vacuity. *British Journal for the Philosophy of Science*, 46:81–110.

Rapaport, A. (1986). *General System Theory.* Abacus Press, Cambridge/ Mass.

Schiffer, S. (1991). Ceteris paribus laws. *Mind*, 100:1–17.

Schurz, G. (1983). *Wissenschaftliche Erklärung. Ansätze zu einer logisch-pragmatischen Wissenschaftstheorie.* dbv-Verlag der TU Graz, Graz.

Schurz, G. (2001a). What is 'normal'? An evolution-theoretic foundation of normic laws and their relation to statistical normality. *Philosophy of Science*, 68:476–497.

Schurz, G. (2001b). Pietroski and Rey on ceteris paribus laws. *British Journal for the Philosophy of Science*, 52:359–370.

Schurz, G. (2002). Ceteris paribus laws: Classification and deconstruction. In J. Earman, Glymour C. and Mitchell, S., editors, *Ceteris Paribus Laws, Erkenntnis* 57:351–372 (special volume).

Schurz, G. and Lambert, K. (1994). Outline of a theory of scientific understanding. *Synthese*, 101:65–120.

Schurz, J. (1990). Prometheus or expert-idiot? Changes in our understanding sciences. *Polymer News*, 15:232–237.

Scriven, M. (1959). Truisms as the grounds for historical explanations. In Gardiner, P., editor, *Theories of History*, pages 443–475, New York/London. The Free Press.

Van Fraassen, B. C. (1989). *Laws and Symmetry.* Clarendon Press, Oxford.

PSYCHOLOGISM, UNIVERSALITY AND THE USE OF LOGIC

Werner Stelzner

1. The place and the task of logic

Philosophers and scientists alike normally distinguish between the so-called rules of nature ('natural rules') and rules of language. Speaking about logic in terms of the rules of logic, we are used to subsume these rules under the rules of language, and not under the rules of nature. These logical rules of language[1] are treated preferably to formulate or to define the deductive space in which the rules of nature are being systematized and to set the formal standards of rational argumentation in science.

Even if one doesn't intend to reduce rationality to deductive-logical rationality, usually a way of performing epistemic acts in the maintenance or in dropping down of epistemic attitudes, is acknowledged to be rational when it is orientated on the observance of logical laws and rules. Accordingly, logical rationality is preferably directed to relations between epistemic attitudes and acts and between their contents, and commonly it does not refer to epistemic attitudes or acts taken as isolated entities. In accordance with this, one can designate persons as rational epistemic subjects, if the establishment of epistemic attitudes and the performance of communicative acts by those persons proceed in agreement with the patterns of logic.

This agreement with the patterns of logic usually has two connected albeit different realizations which are not reducible to each other: One positive component in the sense of active performing or dispositionally admitting the logical consequences of the content of epistemic attitudes or acts, and a second merely negative and passive element demanding the avoidance of contradictions between the contents of epistemic atti-

[1] Cf. the title of an influential German textbook of logic: *Logische Sprachregeln* (Logical rules of language) Sinowjew and Wessel, 1975.

J. Faye, P. Needham, U. Scheffler and M. Urchs, (eds.), Nature's Principles, 269–288.

tudes or acts. Shortly the two components can be referred to by the requirements of logical coherence and logical consistency respectively.

To give in this context of scientific rationality the logical rules a place among the rules of language seems to be in good accordance with the wish to have an absolute standard for rational regulation. This logical standard of rationality should determine that some kind of communicative and mental behavior is formally correct.

Connected with the consideration of logic as part of the rules of language is the idea of undifferentiated absoluteness and universality. In addition, this gives the impression that logic stands outside of all discussions concerning changing standards of rational behavior or cultural-relativized different forms of rationality, i.e., the impression that logical correctness should be the last untouchable bastion of standards for rationality.

This attitude fits in very well with Frege's account of logic: There is only one logic and this logic cannot be differentiated according to varieties in cultural, ethnical or racist respect.[2] So, in fact, the best candidate for a firm untouchable and cultural universal basis for setting up the logical patterns of rules of language seems to be a logic which fulfills the expectations from its objective soundness in the Fregean sense. This soundness is based on the feature of logical rules and logical laws that they do not hold because of epistemic or psychological attitudes towards these rules, but because of their objectivity. This ensures the eternity of such rules and their independence from epistemic attitudes towards these rules. Such attitudes possibly could be changing from time to time, from individual to individual, from social group to social group, from situation to situation, from scientific context to scientific context etc., while the logic is forever the same.

In addition, it is evident that antipsychologistic/objectivistic conceptions about the foundation of logic have the ambition and are claiming to deliver precisely this kind of logic. As an illustration for this, let us take a short look at two prominent advocates of such anti-psychologistic views: Herbart and Frege.

[2]It seems, this should be a hint for people, inclined to misinterpret Frege's remarks concerning Jewish people in his diary ("Tagebuch") from 1924/25 as at least partially unconsciously or instinctively caused by Frege's logical views and his logical ideology, not just as outcome of his political views. Contrary to this opinion, for Frege ethnic or racist different logics are impossible. Even *logics* are impossible, because there is only one logic, independent from any ethnic involvement.

2. The objectivity of logic and anti-psychologism

With the Fregean objectivism closely kindred anti-psychologist views, which were not without influence on Frege's views, were developed by Johann Friedrich Herbart (1776–1841). This cannot be denied even if Herbart's terminology is often connected with psychological notions[3]: "It is true that logic deals with imaginations. But it does not deal with the act of imagining: thus neither with the way and the manner by means of which we arrive at them, nor with the mental state to which we are moved by this." (Herbart, 1851a, 467) Following Herbart, logic is concerned only with this, **what** is imagined, i.e., with the object of imagination.

Herbart (being a forerunner of Frege in this respect) makes a very clear distinction between the soundness (Geltung) and the grasping (Erfassen) of logical laws. And he makes a sharp distinction between the soundness of logic on the one hand and the genesis of theories (or views) about logic on the other hand. Connected with the mentioned objectivity of logical **objects**, Herbart underlines the existence of logical **acts**, which are nevertheless treated as mental acts: "Logical acts: Acts of thinking; which are named judging and concluding. On this occasion, those concepts are always presupposed as already given, and by the composition of those concepts shall arise new concepts." (Herbart, 1851a, 468).

For Herbart, thinking, in which the logical acts are carried out, is only the vehicle in order to grasp the logical. This is followed by the anti-psychologist confession: "Therefore, here too, the logical needs to be kept apart from any intrusion of the psychological." (Cf. Herbart, 1993, 96). With Herbart's uncompromising anti-psychologism is combined a normative and psychologically relevant view on logic, according to which "the whole logic is a **moral** for the thinking, not a **natural history** of the intellect." (Herbart, 1851b, 127).

This characterization of the logic as the moral of thinking points towards applied logic and the relation between logic and sciences. In connection with this, Herbart raises a central question of the whole nineteenth century traditional logic: the relation between pure and applied logic. For Herbart the difference between pure logic and applied logic is treated in the same way as this is later done by Husserl, 1900, where the anti-psychologism is systematically related to pure logic, while applied logic is not free from psychological involvement. Because of this,

[3]Cf. George, 1997, 229–231 about elements of "ungewollten Psychologismus" (unwanted psychologism) with Herbart.

Herbart insists on a strong division between logic and applied logic. Accordingly he demands the "unavoidable sorting out of all psychological, and consequently of the applied logic, which is tangled up in psychological matters" (Herbart, 1993, 81). As to pure logic, developed by Herbart, this had the consequence that it was much criticized, because of its exaggerated purity.

Like Herbart, Frege too accepts a normative function of logic concerning the thinking. But in this prescriptive function of the logical laws, in which "they lay down, how one should think" (Frege, 1903, XV) logical laws have no other psychological status than other laws (e.g., geometrical or physical laws). All these laws can be treated normatively on the ground that they claim soundness and that thinking should proceed in accordance with all this, what holds objectively. Frege sees a dangerous psychologistic threat, actually connected with the speech about the logical laws as laws of thinking. This thread consists in the already by Herbart rejected opinion according to which "these laws would reign over the thinking in the same way and manner, like the laws of nature govern the process in the outer world" (Frege, 1903, XV). In this case the laws of logic would be psychological laws describing the course of the thought process. Consequently, logic would be a branch of (empirical) psychology. However, according to Frege, logic has nothing to do with the laws of psychological processes or states of the counting-for-true (holding for true) or with the laws of psychological acts of judging. Rather, logic is concerned with the laws of being objectively true. And this objectivity is independent from any psychologically relevant epistemic attitudes or states.

Frege finds the cardinal fault of psychologism intertwining of two things, i.e. the taking something for being true and being true itself. It is this confusion which Frege puts in the center of his anti-psychologistic polemics against Benno Erdmann.

Erdmann paved the way for the breaking down of the absoluteness of logical laws: Erdmann treats truth as universal validity in the sense of common acceptance by all the single epistemic subjects. Based on this treatment, Erdmann accepts the possibility of different logics. Frege replies: "Being true is something else than being taken for true, be it by one, be it by many, be it by all, and it is in no way reducible to it." (Frege, 1903, XV)

For Frege, the logical laws are directed to this objective notion of truth: They are the general laws of being true, and as truth does not depend on any psychological facts, a logical law does not depend on psychological laws. And so not only the soundness of logical laws is independent from any epistemic acts or attitudes, logic does not even presuppose the existence of epistemic subjects either. Logic is eternally and universally objectively sound, independently from any empirical situation.

3. The foundation of logic: Psychologism

Despite of all such objectivity, eternity and universality, one may raise not only the question about the place of logic (or, what means to be a logical law or logical rule, or to be *the* logic). Different from such questions about logic, one may raise questions concerning the epistemic grasping of logic and logical rules, as to when and how we can come to know that we have grasped such an objectively, eternally, and universally sound logical law or rule. This concerns the question, how the epistemic relation to logic is founded and from where a historically and empirically given theory takes the license to be that objective logic, which sets the eternally and universally formal patterns of scientific communication and thinking. The question is how one arrives at the "right" logic in order to use it in performing epistemic acts, constructing and proving theories.

Here again we are confronted with the difference between validity (Geltung) and genesis (Genese), or in more Herbartian/Fregean terms, the difference between the soundness of the logical laws and the grasping of these laws by epistemic subjects.

This question cannot be convincingly answered by referring to the objectivity of logic. Now the question would be how one can grasp a logic that is this objective logic: how one can be sure that this logic, which one is actually confronted, is this objective logic. So the question here is not that about the basis for validity (Geltung) of logical laws. The questions is how one can be sure that he (or his social context, or scientific community) has grasped a theory under the name of logic, which in fact deserves this name because of the objectivity of its validity (Geltung). Frege himself was confronted with the above question in his quarrel with formalistic conceptions of logic and arithmetic. Demanding a substantial contentful foundation for formalistic understood theories, Frege himself cannot deliver such a foundation for his own system, because at some definite place he arrives at logically elementary axioms of logic, and the logically elementary (logisch Einfaches) cannot be given a further foundation. They just hold because they are logically elementary. But how one can be convinced of this, i.e., how do I know that the axioms I am facing are objectively sound? Frege here relies on so-called hints (Winke), which may indeed provide at least some help in solving the problem at stake.

As a reaction to the general problem of the foundation of logic, logical psychologism and formalism arose. In some sense formalism is confronted with both objectivism and psychologism. Formalism just refuses to answer the question, because this question seems to have no final answer. Psychologism in the nineteenth century, in fact, in some sense confused soundness (Geltung) and genesis (Genese), or the soundness and the genesis of the grasping of logical laws in thinking. In this respect Frege's antipsychologistic critique is right.

However, the anti-psychologistic critique is not right, if it identifies psychologism with subjectivism, psychologism with empiricism in logic. The following sketch compares some basic attitudes of prominent nineteenth century German psychologistic logicians which cover four scientific generations. And this sketch will demonstrate that the mentioned identification is plainly unjustified. Nevertheless, psychologistic logicians undermined in different ways the claim of absoluteness for logically based rules of language. This not only because of pluralistic conceptions about empirical influenced logics. Such logics would have to consider different features of their area of application or they are changing during the development of mankind. Such logics would depend on the possibly changing psychological constitution of epistemic subjects (e.g., Erdmann, 1892, Sigwart, 1878, Wundt, 1919, Vasil'ev, 1912). Starting from applications of logic in non-logical areas of knowledge, to which psychologistic inclined logicians of the transition period from traditional to modern logic were open-minded, a manifold of non-classical approaches were developed by psychologistic logicians. These non-classical elements made in fact a difference to the body of logic.

However, these non-classical attempts did not become genetic starting points for consecutive developments in the frame of modern logic. And this especially, because in the transition period leading from traditional to modern logic in the result of the anti-psychologistic critique by protagonists of the new logic, psychologism in logic was considered as the main enemy for a well founded logic. As a consequence of the psychologism-quarrel in early 20th century German philosophy, the word "psychologism" became a ideological and pragmatic label for something not worth of serious consideration in developing modern logic.

Another cause for the mentioned neglect of psychological influenced traditional logic lies in the classical orientation of the modern logic that developed according to the Frege-paradigm. Because of its pragmatic orientation, which was characteristic for most of followers of psychologism, the psychologistic line in traditional logic had something threatening for the new logic. From this pragmatic influence threatened the breaking up of the strong classical orientation of the new Fregean logic, which fitted well with the connection between logic and mathematics. All this was connected with the breakdown of the views about objectivity, eternity, and universality of logic. Some kinds of psychologism seemed to pave the way for a manifold of different logics, depending on the difference in the psychological organization of different groups or individuals. Connected with the influence of Darwin's theory and its extension to psychology, psychologism seemed to have opened a way for different logics in different stages of human development.

Notwithstanding the (sometimes unfair) anti-psychologistic critique during the time of the transition from traditional to modern logic, the paradigm of psychologism was strong enough to allow and to promote the development of interesting pragmatically influenced non-classical logical ideas. So, if we look a little closer at the works of psychologistic influenced traditional logicians, we find a manifold of interesting ideas with special relevance to modern non-classical logics. For example, one can point at Entailment theory, the analysis of negation, the so-called logical basic laws of contradiction and excluded middle, the logical treatment of presuppositions, causality, the theory of judgment, propositional acts and attitudes, and last but not least, at the logic of modalities.

4. The diversity of psychologism and the foundation of logic

4.1 Beneke: The genetic-lively (genetisch-lebendige) method in logic

Friedrich Eduard Beneke (1798–1854) with his 1832 *Lehrbuch der Logik oder Kunstlehre des Denkens* is the first eminent of psychologistic philosophers of the nineteenth century, and this in a clearly empiricist style. The main target of his critique is the Herbartian objectivism, which in contrast to Beneke "uncompromising demands that logic should conserve its abstract attitude" (Beneke, 1842, IV)

Beneke's psychologistic attitude towards logic is connected with a general psychologism, which sees in psychology the basic science for all sciences. Sciences are considered as creations of the human mind, of human thinking. If one is interested in the development of a science, one has to bother about laws of psychology first:

> The evolution of the human mind can be influenced only in accord with the law of the mind; and while the ascertainment of these laws belongs to the psychology, in psychology has to be seen the basic science, not only for all other sciences, but for logic too (Beneke, 1842, 17).

In psychology Beneke sees the exclusive basis for the development of a fruitful logic, in order to "develop logic truly as the *Kunstlehre,* the art of thinking" (Beneke, 1842, V). The development of a psychology that underwent a deep reform of its method is the starting point of this endeavor and of the general attraction of psychologism in logic. According to Beneke, psychology is opening the ways for a fruitful development of logic. According to Beneke, the characteristic feature of his own procedure amounts to "all forms and relations of the thinking not to present in abstract form and as completed given in the accomplished soul, but in genetic-lively (genetisch-lebendig) form" (Beneke, 1842, VI). In an epistemic spirit, Beneke demands to consider "not things as such and independently from our imagination and thinking [...], but the things, as they occur in our imagination and thinking, how they are included and processed by the thinking" (Beneke, 1842, 7). Beneke raises the question whether logic should be a science of the real or of the ideal thinking. According to Beneke, logic should include both: the real and the ideal: Logic should not only describe, what really proceeds in the thinking (cf. Beneke, 1842, 8). Logic is the *Gesetzbuch des Denkens*, the "code of thinking", the "law-book of thinking" and the laws and ideals of logic have to be grasped in close connection with the reality of thinking. The ideals of logical thinking have to be natural truths (cf. Beneke, 1842, 9).

Considering this all, we can say that according to Beneke logic is not connected with subjectivism. Psychologism just opens a way to defeat subjectivism in the foundation of logic: Logic this way opens itself for a truly scientific foundation. In this foundation there is no place for accident and arbitrariness: "accident" means just an imperfection of our knowledge. In reality there is no accident, all is governed by strong defined laws of nature. One has to observe and to compare sufficiently precise the facts concerning the natural laws of thinking. Then one can include these laws in the field of our knowledge (Erkenntnis). Consequently, one can control the use of these natural laws in thinking (cf. Beneke, 1842, 16).

4.2 Fries: Anthropological logic

Even if Beneke is recognized as the founder of psychologism, he had – concerning the psychologism in nineteenth century traditional logic – with Jakob Friedrich Fries (1773–1843) an important and often underestimated forerunner. Fries, in opposition to the idealistic followers of Kant, followed Kant in the neglect of the ontologization of logic. On the other hand, he didn't follow the transcendentalism of Kant, but underlined the anthropological-psychological foundation of logic. Already in his foreword to the first edition of his *System der Logik* from 1811, he utters the conviction that "no philosophy and consequently no logic can be **understandably communicated**, if it is not built on anthropologic foundations." (Fries, 1837, IV) This anthropological foundation presupposes the division between demonstrative (formal or philosophical) logic and anthropological logic and the division of logic and grammar. However, the division between anthropological and demonstrative (formal) logic will not constitute the independence of both branches of logic. However, this division is aimed at giving the possibility to work out the anthropological logic in the desired form, in order to get a firm foundation of demonstrative (or formal) logic by anthropological logic.

Fries' presentation of demonstrative (formal) logic as dependent on anthropological logic was criticized by Twesten, Reinhold and Herbart. In the third edition of his *Logic*, Fries reverses this critique against the Herbartian objectivism. Because of the lack of anthropological foundation, Herbart, in Fries' eyes, has not much to say about logic: "With the omission of all psychological, he can teach nothing more about the concepts than saying that they are combinations of attributes (Merkmale)." His own

> anthropological logic has, however, an entirely different point of view. Its main questions are: how concept and thinking come to occur among the activities of the human mind? How they behave towards the remaining acts of knowledge acquisition and how they fit together to the unity of the vivid activity of our mind? [...]
>
> This anthropological logic is worked out automatically intertwined and mixed with all parts of logic. The philosophical [formal] logic is so poor of content and so dependent in all its assertions from the anthropological logic, that one is not able to establish it independently. (Fries, 1837, 3 f.)

The basis for working out the anthropological logic is inner experience. This especially includes inner experience about the rules of language one usually uses in his thinking and concluding. By referring to inner experience, anthropological logic is based in the same manner like empirical psychology of those times. Fries, nevertheless, does not reduce anthropological and philosophical logic to empirical psychology: "However, it would be absurd, to prove by empirical psychology, i. e. by experience, the basic principles of philosophical (demonstrative) logic, the necessary basic laws of the thinkability of things." (Fries, 1837, 5) In this sense, and different from Beneke, Fries in fact is not an foundational psychologist of the empirical kind, but, similar like later Christoph Sigwart, he is merely a psychologist in relation to the genesis, i. e. in the sense of grasping the logical laws. However, all philosophical principles are deducible from anthropological suppositions, which are based on experience. So, Fries discerns between prove, which is based on axioms, on the one hand, and between deducibility, which can be based on empirical facts too, on the other hand. For Fries, psychologism, as it were, opens the way for grasping the axioms and rules of formal logic. These axioms and rules are grasped by deductions from empirical (inner) logical experience. However, this experience does not constitute the laws of formal logic as proved by this experience.

4.3 Sigwart: Psychologism as revealing the logical and norm of thinking

Christoph Sigwart (1830–1904), one of the most eminent traditional psychologistic logicians and one of the main targets of Husserl's anti-psychologistic critique, saw his efforts in the field of logic as precondition for the development of a fruitful applied logic. Sigwart closely connects this orientation on applied logic with the study of logical forms as occurring in colloquial and scientific language.

With his orientation on applied logic, Sigwart differs essentially not only from Herbart, but also from Frege. Freges occupation with logic is aimed at the grasping of the laws and rules of pure logic. And these rules are entirely independent from the area of possible application. For Sigwart, however, logic can fulfil its normative function only then if it is developed in connection with possible areas of application. Sigwart underlines the importance of the study of psychological processes for the grasping of logic, which is fruitfully connected with the theory of methods. Nevertheless, Sigwart sharply delimits his normative conception of logic from descriptive-psychological conceptions of logic. In his view, logic "is not a physics of thinking, but an ethics of thinking" (Sigwart, 1878, 23). This did not prevent him from asserting the "necessity, to find such laws of nature, which govern all judgment" (Sigwart, 1878, 23). Even if logic is the theory of the norms of human thinking, Sigwart finds the way for the grasping of such norms in the study of the area of application of these norms:

> We deny that these norms can be recognized in another way than by studying those natural capacities and functional forms, which shall be ruled by these norms. (Sigwart, 1878, 23)

From the point of view of methods, the judgment as result of a mental act becomes the center of logical interest. Logical categories like truth, logical reason, logical law are directed to the judgment. Contrary to Frege's view, they are not directed to a proposition or to a sentence, which are the possible contents of the judgment.

Nevertheless, Sigwart and Frege remarkable agree both in the accentuation of the hypothetical character of the acknowledgement of logical laws on the one hand and the conclusions drawn from this laws on the other hand. So Sigwart in his *Logik* says:

> There simply cannot be a method to start the thinking from scratch. There can be only a method to continue thinking from given suppositions. In addition, these suppositions, even if acknowledged as uncertain, nevertheless should deliver the starting point for our further thinking. (Sigwart, 1878, 13)

And Frege in *Grundgesetze I* writes:

> The question why and with what right we acknowledge a logical law as true, can be answered by logic exclusively by the reduction of this logical law to other logical laws. Where this reduction is impossible, logic cannot give any answer. (Frege, 1903, XIV)

Frege continues with a clear division between the work of the (objectivistic) logician and the task of the psychologist:

> Leaving the logic, one can say: we are forced to judge by our nature and by the outer circumstances. And if we judge, we cannot deny this law – e.g. of identity, we have to accept it, if we don't wish to confuse our thinking and at the end will resign from any judgement. I will this position neither deny nor confirm it, and will just remark that we here don't have any logical conclusion. Not a reason for being true will be given, but a reason for our holding something for true. (Frege, 1903, XIV)

Frege here criticizes that the psychologistic logician is searching for logical reasons, where logic has to be silent. In a similar sense, Lotze acknowledges the existence of non-logical reasons for logical laws, and logic has to be silent about such reasons. This occurs where reasons for the acceptance of logical principles are searched in the realm of psychology. When Frege mentioned that in order to find out if one is forced to judge one has to leave logic, then, according to Sigwart, this leaving of the logic belongs to the business of the logician. Here is opened a field of activity for the psychologistic logician. And this field is closed for logicist Frege, who remains in the realm of objective valid logic. A final foundation (Letztbegründung) for the logical rules neither Sigwart finds when he aborts the foundational chain. However, he comes to a pragmatic closure, which shows itself as a principle of logically foundational charity, based on inner evidence:

> The possibility to establish the criteria and rules of the necessary and universally sound principles of the progress in thinking rests on the ability to discern objective necessary thinking from the not necessary thinking. And this ability manifests itself in the direct consciousness of evidence, which accompanies necessary thinking. The experience of this consciousness and the belief in its reliability is a *Postulate*, which cannot be avoided. (Sigwart, 1878, 15 f.)

In Sigwart's understanding, logic has to fulfill its task in the area of the logical reasons for knowledge. For this, the objectivity of the logical reasons is unavoidable. In order to become logical reasons, however, these reasons (in contrast to Frege's attitude) have to be grasped: logical reasons, which we don't know, in Sigwart's understanding are just an absurdity, an *Unding*, a contradiction in itself.

Notwithstanding the connection between logic and psychology, in his theory of modalities Sigwart develops a view of logical necessity, which is entirely founded on basing such a necessity on rules of language.

Logical necessities are based exclusively on the meaning of the occurring words. Therefore, logical necessities are speaking about our terminology. Because of this, they are empirically neither falsifiable nor verifiable. The tight connection between logical necessity and rules of language once more makes understandable Sigwart's prevalent interest in the logical analysis of language.

4.4 Wundt: Normative psychological logic as empirical science sui generis

The close connection between logic and different branches of science is a basic feature of the logical enterprise of Sigwart's contemporary Wilhelm Wundt (1832–1920), the founder of empirical psychology. However, one should not conclude that Wundt would be an empirical psychologistic logician. He, like Sigwart, underlines the normative function, which can be adequately and fruitfully fulfilled only then, when one considers specific features of the areas of sciences where logic has to be applied.

Wundt's main interest in logic is determined by his aim to develop logic as applied logic. This applied logic should specify the logical rules of language for different branches of sciences, ranging from mathematics to social sciences.

In Wundt's treatment of logic, there is a common part of logic in all the different specific logics of sciences, and this common part is the general logic. However, even this general logic is not to be understood as developed independently from the fields of applications of the specific applied scientific logics. Wundt is convinced that in order to fulfill its normative task, logic should not treat scientific knowledge from the outside, like an isolated and independent alien:

> If the logic is determined to be submitted to those conditions to which the science is submitted everywhere, then, logic cannot proceed under the condition that the forms of thinking would be indifferent in relation to the content of knowledge. (Wundt, 1919, 8)

Logic has to behave like other citizen of the realm of science, even if it has special tasks. Starting from scientific practice, Wundt aims at an adequate analysis of the logical relations inhering in the sciences. So for Wundt a view on logic as part of predefined and presupposed rules of language apart from scientific content would be unacceptable. In order to get the logical rules of language, applicable for a special science, you have to take into account this science. Again, like for Sigwart, for Wundt from this results a kind of psychologism concerning the genesis of logic. However, it is the normative logical psychologism, which in Wundt's opinion on the one side connects logic with psychology, and on the other side discerns it from psychology. Wundt expresses this very clearly in his rejection against the accusation of psychologism, which Husserl (Husserl, 1900) raised against Wundt in the first volume of *Logical investigations*. In Wundt's opinion, the accusation is false, because logic and psychology are divided by prescriptivity vs. descriptivity concerning the thinking:

> While psychology is teaching us how the course of our thought really proceeds, logic aims to determine how it should proceed in order to lead to scientific knowledge. (Wundt, 1919, 1).

However, even this division does not by itself prevent elements of logical psychologism from playing a role in the foundation of logic. In order to make clear that his logic is not a kind of psychology, Wundt admits that logic fulfills its normative task as an empirical science, as an empirical science of special kind to be sure:

> This work [Wundt's *Logik*] aims to be something entirely different from a psychology of thinking [...]. It aims to be an empirical science *sui generis*, what psychology – according to its real content and according to its suppositions – never can be. (Wundt, 1919, VIII)

Despite of his differentiation between logic and psychology, Wundt's logic is not independent from psychology. Wundt, nevertheless, believes in the "inseparable connection (Gebundenheit) of the logical laws to the psychological forms of the development of thinking." (Wundt, 1919, 90). Therefore, even if Wundt divides logic from empirical psychology, a kind of transcendental logical psychologism is apparent.

4.5 Erdmann: Implicit foundational psychologism

Benno Erdmann (1851–1921), like Sigwart, Wundt, and other psychologistic logicians, reduces truth to the acknowledgement of something as true (i.e., sound, correct etc.). Nevertheless, he does not consider himself as a defender of validity-theoretical foundational psychologism: This kind of psychologism derives the soundness of logical laws from the psychologically determined genesis of the acceptance of these laws. Contrary to this, Erdmann claims to defend simply a normative psychologism:

> The judgements, conclusions, definitions, partitions etc. are processes of the consciousness, which stand in lawful connection to each other and to other processes of the imagination and of the sensation and the will. From this, however, does not follow that the object of logic would be an object of psychology [...] Logic does not investigate the processes related to the actual conditions of their origin, their development and connections. Logic remains with the question, how they should be in order to become universally valid propositions about the imagined. (Erdmann, 1892, 18).

Looking at Erdmann's statement, one should remember the special sense in which universal validity is introduced by Erdmann as definition and criteria for the soundness of logical laws. By Erdmann and other advocates of normative psychologism like Sigwart and Wundt universal validity is not objectively defined, like in the Fregean objectivistic anti-psychologistic treatment of universal validity, but is taken as the expression of common agreement or common agreeableness. A judgement is universally sound,

> if its object, i.e., what is imagined in subject and predicate, for all is the same, objective and commonly certain, and the statement about the object, which is performed by the judgment, is necessary to be thought. Consequently, the necessity to think is objective, it follows from the conditions of our thinking in accordance with the nature of its objects. (Erdmann, 1892, 6)

This way, the validity of logical laws and logical rules of language is based on psychological criteria and this kind of normative psychologism comes out as implicit foundational psychologism. This kind of implicit foundational psychologism is connected with the abundance of the intangibility and eternity of logical principles. These principles depend on the psychological organization of the epistemic subjects. The talk of the "necessity for the thinking" (Denknotwendigkeit) then is relativized to the given empirical situation, in which the thinking takes place:

> Our logical principles also in view of this retain their necessity for the thinking. However, this necessity is seen not as absolute, but as hypothetical. We cannot act in another way than to agree with these principles – according to the nature of our imagining and thinking. They hold generally, presupposed, that our thinking remains the same. (Erdmann, 1892, 378)

Following this, even logical rules of language are changed with changing conditions for the thinking itself. Therefore, what falls down with this kind of implicit foundational psychologism is the universality and the eternity of logical rules of language. In this sense, then, logical rules of language have no other status then other descriptive or conventional rules of language, which can change from time to time or from one group of language-users to another group.

4.6 Lipps: Explicit foundational psychologism: Logic as physics of thinking

Theodor Lipps (1851–1914) gives in his *Logik* from 1893 an explicit confession of foundational psychologism and argues explicitly against any normative treatment of logic. Logic is the science about the forms and laws of thinking, and it is unnecessary to characterize logic as "science of the normative laws of thinking, or as science how one should think in the right way", "because we are thinking correct in every case if we are thinking". (Lipps, 1893, 1). While Erdmann, Wundt and Sigwart considered logic as the "Ethics of thinking" which is not a "physics of thinking", Lipps takes the converse line: According to Lipps, logic is the **physics of thinking** and only on this basis logic is a norm of thinking, because "the question what one should do, is in every case reducible to the question what one should do in order to attain a certain aim and this is equivalent with the question how one can actually attain this aim" (Lipps, 1893, 1).

The laws of thinking referred to by Lipps in his definition of the subject of logic are treated as psychological laws. And following this,

> logic is a psychological discipline. This is just as certain as the cognition occurs only in the psyche and as the thinking is a psychical event which completes itself in the cognition. (Lipps, 1893, 1 f.).

As a special discipline of psychology, the logic distinguishes itself by the investigation of the relation between knowledge and error.

If seen in the perspective of logical rules of language, in Lipps' understanding, however, logical rules of language are just the empirical given habits to use a language, no special normative content is connected with these "logical" rules of language.

5. Psychologism and the use of logic in epistemic contexts

I hope the above discussion has illustrated some aspects of the extensive diversity that obviously is not only ruling between psychologism and antipsychologism on the one hand, but as a matter of fact is playing a role even within psychologism too. Where its influence is felt within psychologism, it concerns questions about the foundation of logic as a basis for the explication of adequate and rationally founded and universally valid logical rules of language, suitable for application in science. A clear-cut classification of logicians into psychologistic and antipsychologistic at least because of the huge variety of the so-called psychologism in logic seems not to be appropriate here. Such a clear-cut division merely is obscuring the problem to develop a foundation for such logical rules of language.

The picture of the division between the good and the bad guys, which are subsumed under the label of antipsychologism on the one side and psychologism on the other side, may have some ideological value. In view of the differentiated development and use of logical resources in science and pragmatic contexts, there is no substantial value of undifferentiated focussing on this division. The remarkable thing about the psychological attitude of philosophers such as Sigwart and Wundt consist in their orientation on the occurrence of differentiated logical non-classical structures which occur in the given languages of science and colloquial contexts. Here we find their inclination to approach such logically different structures as factually given. While pursuing an empirical approach to the application of logic, these logicians do not define away the diversity of non-classical logical structures. The main reason for this is that they don't follow a predefined conception of objectively given classical logical structure.

In view of nowadays variety of logics, the question of the use of logic, which out of a dozen of logics is to be use for the establishment of logical rules of language, cannot be answered by means of the so-called objectivity of these logics only. For what does objectivity mean in the face of this variety? In a sense, all these differently developed logics can be called objective, whatever the epistemic or psychological attitude towards them is. For they are formal and often even formalizations of some segment of language, of some part of metaphysics, or of some philosophy of mind.

Insofar, every logic is universal. However, this way of talking about the universality of logic does not say very much. This talk is not false, but vacuous. For instance, in the above-mentioned textbook *Sprachregeln* a convincing argumentation for the universality of logic is developed followed by the development and presentation of at least ten different kinds of logical entailment.

Of course, logic is universally sound, and so are the logical rules of language established by these logics. This seems to be relieving, but unfortunately, there are so much of these universally sound logics and different kinds of logical rules of language. Which of these logical rules of language one should use, i. e., which is the fitting logic for a given context applicable for a given task? Which logically determined set of rules of language is theoretical adequate and fruitful in order to be used for the given task. Just to mention one example: rational epistemic logic is undoubtedly universally valid, in some interpretations adequate, but there can be doubts about its fruitfulness as an epistemic logic: Because of the theoretical construction of such a logic no fitting empirical epistemic subject can be found. Therefore, in fact not universality and abstract soundness is required. Rather, special applicability and fruitfulness, related to the supposed context of use for this logic, should be the criteria for choosing the right logic. And at this point, some kind of psychologism can intrude itself again. Scholz/Hasenjaeger can be right in saying, "Which logic one chooses depend on the fact, what kind of human one is". This is just a description of a fact. However, if the question is what kind of logical rules of language one should choose, in order to decide rationally and to come to sound conclusions, they are wrong. Here much depends on the area of application of these rules and of the direction of application, i.e., what is my aim with this application, is it, e.g. my aim to convince someone that I am bright or that someone should do this and this.

To decide these questions one has to consider the empirically given situation in a highly complex sense: This includes the status and the genesis of fitting terminologies, questions of the use of language in relation to which the possibly applied logical laws and rules of language should be adequate. In order to mention some problems to be taken into account here:

> Differences between *languages for analyzing and describing a situation* and *languages successfully used in a situation* by epistemic subjects;
>
> The existence of different types of epistemic subjects concerning their logical abilities in the application of rules of language;
>
> Different logical capacities of epistemic subjects;
>
> The possibility of special entailment relations in epistemic contexts;
>
> Should one use epistemic sensitive or insensitive formalizations and entailment relations?
>
> Should the standards of analysis be set by theoretically inclined epistemic considerations only or should there be an empirical component in these considerations?

Etc.

Monistic conceptions of logical rationality and of the universality of logical rules of language ignore, be it consciously or unconsciously, that the claim of absoluteness of logic is doubted not only from outside logic, but that the development of logic itself makes it extremely difficult to maintain such a claim of absoluteness concerning (absolute) logical rationality. Not only the present manifold of different developed non-classical logics, which partly stood in mutual competition against each other and especially against classical logic, contributed to this situation. To make this situation clear contributed essentially the different foundational conceptions concerning logic, especially the differences between differentiated psychologistic and anti-psychologistic conceptions. Especially in connection with the application of logical rules of language in science and colloquial contexts, this all relates directly to such questions as about the liabilities, reliability and availability of logical principles and appropriate rules of language as standards for (logical) rationality of science and in different other epistemic contexts.

Acknowledgements. The author is grateful to the *German Research Foundation* (*Deutsche Forschungsgemeinschaft*, DFG – Bonn) and the *University of Bremen* for their invaluable support, which made it possible to realize this research.

References

Beneke, F. (1832). *Lehrbuch der Logik als Kunstlehre des Denkens.* Berlin/Posen/Bromberg.

Beneke, F. (1842). *System der Logik als Kunstlehre des Denkens, Erster Theil.* Ferdinand Dümmler, Berlin.

Bochenski, I. (1956). *Formale Logik.* Alber, Freiburg i. B. and München.

Boole, G. (1847). *The Mathematical Analysis of Logic. Being an Essay Towards a Calculus of Deductive Reasoning.* Cambridge/London.

Boole, G. (1854). *An Investigation of the Laws of Thought, on which are Founded the Mathematical Theories of Logic and Probabilities.* London.

Erdmann, B. (1892). *Logik, 1. Band: Logische Elementarlehre.* Max Niemeyer, Halle.

Frege, G. (1879). *Begriffsschrift, eine der mathematischen nachgebildete Formelsprache des reinen Denkens.* L. Nebert, Halle a. S.

Frege, G. (1884). *Grundlagen der Arithmetik.* W. Koebner, Breslau.

Frege, G. (1893/1903). *Die Grundgesetze der Arithmetik*, volume I, II. H. Pohle, Jena.

Fries, J. (1837). *System der Logik.* Christian Friedrich Winter, Heidelberg.

George, R. (1997). Psychologism in logic: Bacon to Bolzano. *Philosophy and Rhetoric*, 30:213–242.

Herbart, J. (1850–1852). *Sämmtliche Werke*, volume 1–12. Leopold Voss, Leipzig.

Herbart, J. (1851a). *Hauptpunkte der Logik.* In Herbart, 1852, vol. 1, pp. 466–478.

Herbart, J. (1851b). *Lehrbuch der Psychologie.* In Herbart, 1852, vol. 5, pp. 1–190.

Herbart, J. (1993). *Lehrbuch zur Einleitung in die Philosophie.* Felix Meiner, Hamburg.

Husserl, E. (1900). *Logische Untersuchungen*, volume I. C. E. Pfeffer (Robert Stricker), Halle a. S. (Meiner, Hamburg 1992).

Kneale, W. and Kneale, M. (1962). *The Development of Logic.* Oxford University Press, Oxford.

Kusch, M. (1995). *Psychologism.* Routledge, London/New York.

Lipps, T. (1893). *Grundzüge der Logik.* Voss, Hamburg.

Lotze, R. (1843). *Logik.* Weidmann'sche Buchhandlung, Leipzig.

Lotze, R. (1874). *Logik. Drei Bücher vom Denken, vom Untersuchen und vom Erkennen.* S. Hirzel, Leipzig. (= Lotze, *System der Philosophie*, T. 1). New edition: Lotze, Rudolph Hermann, 1989,*Logik. Erstes Buch. Vom Denken (Reine Logik).* Philosophische Bibliothek Band 421, Meiner, Hamburg. Lotze, Rudolph Hermann, 1989, *Logik. Drittes Buch. Vom Erkennen (Methodologie).* Philosophische Bibliothek Band 408, Meiner, Hamburg.

Peckhaus, V. (1997). *Logik, Mathesis universalis und allgemeine Wissenschaft.* Akademie-Verlag, Berlin.

Rath, M. (1994). *Der Psychologismusstreit in der deutschen Philosophie.* Alber, Freiburg/München.

Schröder, E. (1877). *Der Operationskreis der Logik.* Leipzig.

Schröder, E. (1909/1910). *Abriß der Algebra der Logik*, volume I and II. Leipzig/Berlin and Leipzig. I: Elementarlehre, II: Aussagentheorie, Funktionen, Gleichungen und Ungleichungen.

Sigwart, C. (1873/1878). *Logik*, volume I and II. Heinrich Laupp, Tübingen. I: Die Lehre vom Urteil, vom Begriff und vom Schluss; II: Die Methodenlehre (2. Aufl. Freiburg 1889–1893, 4., durchgesehene Auflage, J. C. B. Mohr (Paul Siebeck), Tübingen 1911).

Sinowjew, A. and Wessel, H. (1975). *Logische Sprachregeln.* Deutscher Verlag der Wissenschaften, Berlin.

Vasil'ev, N. A. (1910). On particular judgments, the triangle of opposites, and the law of excluded fourth. *Uchennye zap. Kazan. un-ta*, 77, kn. 10:1–47. (Васильев, Н. А.: О частных суждениях, о треугольнике противоположностей, о законе исключенного четвертого. *Ученные зап. Казан. ун-та* 77, кн. 10, pp. 1–47).

Vasil'ev, N. A. (1912). Imaginary (nonaristotelian) logic (russ.). *Zhurnal m-va nar. prosveshcheniya*, Nov. ser. (Ch. S. 1912, Ch. 40):207–246. (Васильев, Н. А.: Воображаемая (неаристотелева) логика. In: *Журнал м-ва нар. просвещения Нов. сер.* (1912, Ч. 40), 1912, pp. 207–246).

Windelband, W. (1913). Die Prinzipien der Logik. In: *Encyclopädie der philo-sophischen Wissenschaften*, in Verbindung mit Wilhelm Windelband herausgegeben von Arnold Ruge, Verlag von J. C. B. Mohr (Paul Siebeck), Tübingen, pp. 1–60.

Wundt, W. (1880/1919). *Logik. Eine Untersuchung der Prinzipien der Erkenntnis und der Methoden wissenschaftlicher Forschung*, volume I and II. Stuttgart. I: Allgemeine Logik und Erkenntnistheorie (1880, 1893, 1906, 1919), II: Logik der exakten Wissenschaften (1883, 1894, 1907, 1920).

Index

GPSR Compliance

The European Union's (EU) General Product Safety Regulation (GPSR) is a set of rules that requires consumer products to be safe and our obligations to ensure this.

If you have any concerns about our products, you can contact us on ProductSafety@springernature.com

In case Publisher is established outside the EU, the EU authorized representative is:

Springer Nature Customer Service Center GmbH
Europaplatz 3
69115 Heidelberg, Germany

Batch number: 10755284

Printed by Printforce, the Netherlands